THE MATHEMATICAL WORK

OF

JOHN WALLIS

Reproduced, by permission, from the Original in the Possession of the Bodleian Library, Oxford.

THE MATHEMATICAL WORK

OF

JOHN WALLIS,
D.D., F.R.S.
(1616–1703).

BY

J. F. SCOTT, Ph.D., B.A.

With a Foreword by
E. N. da C. Andrade, D.Sc., Ph.D., F.R.S.,
Quain Professor of Physics in the University of London.

CHELSEA PUBLISHING COMPANY
NEW YORK, N. Y.

SECOND EDITION

THE PRESENT WORK IS PUBLISHED AT NEW YORK, N. Y. IN 1981. IT IS A SECOND (TEXTUALLY UNALTERED) EDITION OF THE BOOK THE MATHEMATICAL WORK OF JOHN WALLIS, BY JOSEPH FREDERICK SCOTT, WHICH WAS ORIGINALLY PUBLISHED AT LONDON IN 1938. THIS BOOK IS PRINTED ON 'LONG-LIFE' ACID-FREE PAPER.

INTERNATIONAL STANDARD BOOK NUMBER 0-8284-0314-7

LIBRARY OF CONGRESS CATALOG CARD NUMBER 80-85524

PRINTED IN THE UNITED STATES OF AMERICA

CONTENTS.

	Page
FOREWORD	v
PREFACE	vii
CHAPTER I. Early Life and Training	1
II. The Beginnings of the Royal Society. Wallis's Part in its Early History. Appointment as Savilian Professor of Geometry	8
III. The Treatise on the Conic Sections	15
IV. The *Arithmetica Infinitorum*	26
V. The *Mathesis Universalis*. The *Commercium Epistolicum* (1657/8). Controversy with Fermat and other Members of the French Mathematical School	65
VI. Appointment as *Custos Archivorum*. The Quarrel with Doctor Holder	83
VII. The *Mechanica, sive Tractatus De Motu*	91
VIII. The Hooke-Hevelius Dispute. Publication of Ancient Manuscripts	127
IX. The *Treatise of Algebra*	133
X. The Dispute with Hobbes summarised	166
XI. Conclusion. Importance of Wallis's Work to succeeding Generations	173
APPENDIX I. Brief Biographies of Wallis's Contemporaries and Immediate Predecessors, who are mentioned in this Work	180
II. Some Observations on the Development of Notation during the Seventeenth Century, with Specimens of Notation then Current	218
III. List of Wallis's Mathematical Works, including his Contributions to the *Transactions*	230
BIBLIOGRAPHY	234
INDEX	235

"Et sunt quatuor scientiæ magnæ, sine quibus cæteræ scientiæ sciri non possunt nec rerum notitia haberi . . . . Et harum scientiarum porta et clavis est Mathematica."

ROGER BACON, *Opus Majus. Pars Quarta.*

FOREWORD.

By Professor E. N. DA C. ANDRADE, D.Sc., Ph.D., F.R.S.

MOST of the current general histories of science contain grave faults both of fact and of interpretation, due to the circumstance that their statements are too often based on a study of earlier writers on the subject, who themselves have uncritically accepted corrupt tradition or have served partial ends, rather than on a study of the works of the great pioneers. It is a fortunate feature of the present time that a number of scholars have realized the importance and interest of following the growth of the ideas that lie at the basis of our modern mechanical civilization and of, in many cases, our philosophical attitude to nature. These scholars, often devoting to their studies the whole of the scanty leisure which is left them by their professional pursuits, are enthusiastically investigating the performances of the great men of science, as seen against the background of the general history of connected learning and of contemporary events, and are rapidly accumulating reliable material for a really informed history of science. It is a pleasant privilege to welcome Dr. J. F. Scott into this band of new historians who, serving no ideology—unpleasant words fitly denote unpleasant phenomena—are patiently endeavouring to remove the obscurity effected by learned language and forgotten terminologies, to see with the eyes of other times and to assess the contributions of original minds to our inherited store of scientific knowledge.

John Wallis is one of the most interesting figures in that great period which saw the birth of the Royal Society and the rapid development of the conceptions that lie at the basis of modern science. He was not only a great innovator in mathematics, but a man who, by his personal activities and influence, did much to inspire his contemporaries with his own enthusiasm for the advancement of scientific truth. He was also a brilliant and witty controversialist: Dr. Scott deplores the long quarrel with Thomas Hobbes, but Wallis had great provocation, and whoever has read *Hobbius Heautontimorumenos* will find it hard to regret that he took up the challenge. He had his faults—in particular his ungenerous, often unjust, attitude to the French school of mathematicians is a reproach to his character. These faults Dr. Scott does not attempt to palliate. The picture which he gives us is a balanced and understanding one, based upon patient and exhaustive study, and when he delivers judgment he always provides the evidence on which his reader can, if he wishes, form his own verdict.

FOREWORD

I am particularly struck with the care with which Dr. Scott has analyzed Wallis's achievement in pure mathematics, and the knowledge which he shows of the contemporary background. The account of Wallis's discussion of impact is equally good. The book has, further, the great merit of readability : the narrative is clear and straightforward, even when it is dealing with highly technical matters. Dr. Scott has not only achieved a detailed understanding of all aspects of Wallis's extensive work, but has ordered his account admirably. His work will be indispensable to those interested in the early history of the Royal Society. I commend to all students of the seventeenth century, whether scientific or humane, this learned and lucid book.

E. N. DA C. ANDRADE.

PREFACE.

THIS work aims at giving an account of the contributions of John Wallis to the development of mathematics during the latter half of the seventeenth century.

Despite the reputation he enjoyed amongst his contemporaries, despite the extensive works he has left behind him, Wallis has never received his full share of recognition from the historian of science. In the pages which follow are suggested reasons for his comparative obscurity, and an attempt is made to re-establish his reputation as a mathematician and as a precursor of the mighty Newton.

The seventeenth century was a period of profound political unrest, but the spirit of investigation and learning was kept alive by a few zealous enthusiasts, among whom Wallis was conspicuous. The work draws attention to his part in the formation of the Royal Society, which during that period became the focus of scientific culture in this country, and which attracted the attention of some of the most distinguished men of science on the Continent.

Wallis's title to fame rests on the influence he exercised on his contemporaries through his correspondence—much of which was of a controversial nature—and on his published works, which proved an inspiration to his immediate successors, and even to-day are worthy of study as monuments of scholarship and ingenuity. This work contains a summary of his principal mathematical treatises in modern form. In the *Arithmetica Infinitorum* the steps which led to his famous π series are traced in detail. In this Chapter is indicated the extent of Newton's indebtedness to Wallis. Attention is called to the treatise on the *Conic Sections* which preceded the *Arithmetica Infinitorum*. By investigating the properties of these curves with the aid of the powerful Cartesian analysis he revived interest in this important branch of mathematical science; moreover, by his free use of the index notation he did more to jettison the old-fashioned symbolism than any other writer of his day.

In the Chapter which deals with the treatise *De Motu*, Wallis's contributions to the hitherto neglected science of mechanics are described. This is shown to be a work of profound scholarship, and one which vividly illustrates the chaotic state of the science during the first half of the seventeenth century. The extent to which Wallis was instrumental in stimulating interest in it and in giving it coherence is discussed.

In the early chapters of the *Treatise of Algebra* Wallis shows an intimate acquaintance with the history of the subject. Its development from antiquity to the Middle Ages is traced by Wallis with a fullness and a precision hitherto unknown. The account of the history of mathematics in his own century is, however, completely vitiated by the grossest partiality towards the English mathematical school and blindness towards the achievements of the French. It is suggested that this attitude, which Wallis so persistently adopted, and which is reflected again and again in his correspondence, was not without influence on that unhappy alienation of the English and Continental mathematical schools which was such a distressing feature of eighteenth century mathematics.

Finally, his mathematical work is summarized, his outstanding contributions being emphasized.

Three Appendices are subjoined :—

I. Brief Biographies of Wallis's contemporaries and immediate predecessors who are mentioned in these pages.

II. Some Observations on the Development of Notation during the Seventeenth Century, together with Specimens of Notation then current, *e. g.* :—
    i. Oughtred, *Clavis Mathematicæ* (1631).
    ii. Harriot, *Artis Analyticæ Praxis* (1631).
    iii. Wallis, *De Sectionibus Conicis* (1655).
    iv. Wallis, *Arithmetica Infinitorum* (1656).
    v. Wallis, *Algebra* (1685).

III. List of Wallis's mathematical works, including his Contributions to the *Philosophical Transactions*.

In addition to all Wallis's works mentioned in Appendix III the following Treatises have been consulted. They are referred to in the footnotes by the brief title.

Title of work.	Brief title.
AUBREY, JOHN. Lives of Eminent Men. 2 vols. London, 1813.	Aubrey.
BAILLET, A. La Vie de M. Descartes. Paris, 1691.	Baillet.
BALIANI. De Motu Naturali Gravium Solidorum et Liquidorum. Genoa, 1638 and 1646.	Baliani.
BALL, W. R. R. A Short Account of the History of Mathematics. Cambridge, 1889.	Ball.

PREFACE

Title of work.	Brief title.
BALL, W. R. R. History of the Study of Mathematics at Cambridge. Cambridge, 1889.	Ball (*Camb.*).
Biographia Britannica. London, 1763.	B. B.
Biographie Universelle. Paris, 1843–1858.	Biog. Univ.
BIOT et LEFORT. L'Analyse Promota. Paris, 1856.	Biot et Lefort.
BIRCH, THOMAS. History of the Royal Society of London. 4 vols. London, 1756/57.	Birch, *Hist. Roy. Soc.*
BIRCH, THOMAS. The Works of the Hon. Robert Boyle. 5 vols. London, 1744.	Birch (1774).
Bulletin of the American Mathematical Society, Vol. VII. 1901.	Amer. Math. Soc.
Bullettino di Bibliografia e di Storia delle Scienze Matematiche e Fisiche. B. Boncompagni. Rome, 1868–87.	Boncompagni.
BREWSTER, DAVID. The Life of Sir Isaac Newton. London, 1831.	Brewster.
CAJORI, FLORIAN. History of Mathematics. New York and London, 1894.	Cajori.
CAJORI, FLORIAN. History of Mathematical Notations. 2 vols. Chicago, 1928.	Cajori (*Notations*).
CANTOR, MORITZ. Vorlesungen über Geschichte der Mathematik. Leipzig, 1894–1908.	Cantor.
CHASLES, M. Aperçu Historique des Méthodes en Géométrie. Paris, 1875.	Chasles.
CLARK, GILBERT. Oughtredus Explicatus, sive Commentarius in Ejus Clavem Mathematicam. London, 1682.	Clark, G.
COLLINS, JOHN. Commercium Epistolicum. London, 1712.	Com. Epist.
DANNEMANN, F. Die Naturwissenschaften in ihrer Entwicklung und in ihrem Zusammenhange. Leipzig, 1911.	Dannemann.
DE MORGAN, A. Essays in the Life of Newton. (Ed. Jourdain.) London, 1914.	De Morgan.
DESCARTES, RENÉ. Lettres (3 vols.). Paris, 1657/67.	Descartes (*Lettres*).
DESCARTES, RENÉ. Principia Philosophia, 3rd Edn. Amsterdam, 1656.	Descartes (*Prin. Phil.*).
Dictionary of National Biography.	D. N. B.
EDLESTON, J. Correspondence of Newton and Cotes. London, 1850.	Edleston.
Encyclopædia Britannica, xivth edition.	E. B.
EVELYN, JOHN. Diary. (Bray's edition.)	Evelyn.
FERMAT, PIERRE. Varia Opera Mathematica. Tolosæ, 1679.	Fermat.
GIRARD, ALBERT. Invention Nouvelle en l'Algèbre. Reimpression, Leiden, 1884.	Girard.

Title of work.	Brief title.
HARRIOT, THOMAS. Artis Analyticæ Praxis. London, 1631.	Harriot.
HEARNE, THOS. An Account of some Passages in His Life, addressed to Dr. Thos. Smith, dated Oxon. Jan. 29, 1696/7. It is published in the Appendix to the Preface by the Editor, Thos. Hearne, to his Edition of Peter Langtoft's Chronicles. Oxford, 1725.	Hearne.
HOLDER, Dr. THOS. A Supplement to the Phil. Trans., July 1670, with Some Reflexions on Dr. John Wallis, his Letter there Inserted. London, 1678.	Holder.
HOOKE. Diary of. Ed. Robinson and Adams. London, 1935.	Hooke Diary.
HUTTON, C. Mathematical Dictionary. 1815.	Math. Dict.
HUYGENS, C. Œuvres. La Haye, 1888–1934.	Huygens.
LAGRANGE. Mécanique Analytique. Paris, 1788.	Lagrange.
MACH, ERNST. Science of Mechanics. Chicago, 1907.	Mach.
Mathematical Gazette, Dec. 1910 and Jan. 1911. Article on Wallis by Sir T. Percy Nunn.	Math. Gazette.
MOLESWORTH, Sir WILLIAM. The English Works of Thos. Hobbes. London, 1839.	Molesworth.
MONTUCLA, J. F. Histoire des Mathématiques, 1799–1802.	Montucla.
Nouvelle Biographie Générale. (Hoefer, 1853–7.)	Nouv. Biog. Gén.
ORNSTEIN, M. The Rôle of Scientific Societies in the xviith century. Chicago Univ. Press, 1928.	Ornstein.
OUGHTRED, WM. Clavis Mathematicæ. London, 1631.	Oughtred.
PASCAL, BLAISE. Récit de la Grande Expérience de l'Équilibre des Liqueurs. Paris, 1648.	Pascal.
PEPY'S DIARY. (Braybrooke's edn.) 1911.	Pepys.
Philosophical Transactions.	Phil. Trans.
POGGENDORFF. Biographisch-literarisches Handwörterbuch zur Geschichte der Exacten Wissenschaften. Leipzig, 1863–1904.	Poggendorff.
Record of the Royal Society. Third edition, 1912.	Record, R.S.
RIGAUD, S. Correspondence of Scientific Men of the xviith Century. Oxford, 1841.	Rigaud.
STEVIN, S. Œuvres. Leyden, 1634.	Stevin.
SPRAT, THOMAS. The History of the Royal Society of London, for the Improvement of Natural Knowledge. 1st Edn., London, 1667.	Sprat.
THOMSON, THOMAS. History of the Royal Society from its Institution to the end of the xviith Century. London, 1812.	Thomson.

PREFACE

Title of work.	Brief title.
WALLIS, JOHN. A Defence of the Royal Society. In Answer to the Cavils of Dr. William Holder. London, 1678.	Wallis, *Defence R. S.*
WARD, J. Lives of the Professors of Gresham College. London, 1740.	Ward.
WATT, R. Bibliotheca Britannica. Edinburgh, 1824.	Bibl. Brit.
WELD, C. R. History of the Royal Society. 2 vols. London, 1848.	Weld.
WHEWELL, W. History of the Inductive Sciences, 3rd edition, 1857.	Whewell.
WOOD, A. Athenæ Oxonienses and Fasti Oxonienses. London, 1691/2.	Wood.

The author gratefully acknowledges his indebtedness to :

(1) Professor E. N. da C. Andrade, D.Sc., Ph.D., F.R.S., for having so kindly written the Foreword.

(2) Dr. A. Ritchie-Scott, F.R.S.E., Principal of the L.C.C. Beaufoy Institute, for his kindness in reading the proofs, and for many helpful suggestions.

(3) Mr. H. W. Robinson, Librarian of the Royal Society, and Mr. J. C. Graddon, A.R.C.S., Assistant Librarian of the Society, both of whom have placed their wide knowledge of the History of Science at the disposal of the author.

(4) Mr. T. L. Wren, M.A., of the Department of Mathematics, University College, London, for a great deal of help and useful criticism.

The author also wishes to place on record his great thanks to the University of London Publications Fund for their generous grant towards the publication of this work ; and to the Council of the Royal Society, and to the Bodleian Library, Oxford, for permission to consult manuscripts in their possession ; also to Messrs. Taylor & Francis, Ltd., for the care and skill which they have devoted to the production of the book.

J. F. SCOTT.

London, 1938.

CHAPTER I.

Early Life and Training.

Though more than two centuries have elapsed since the death of John Wallis, the historian of science has so far paid but scant tribute to the memory of this remarkable man. It is not easy to understand why a mere pittance of immortality should have been accorded to one who in his day occupied such an exalted position in the world of science. For it seems beyond all doubt that amongst his contemporaries at least Wallis was esteemed eminent, not only in his adopted profession of Divinity, but also over a range of subjects which was truly encyclopædic. Mathematics, mechanics, sound, philology, the phenomena of the tides, even music—in all these his writings give evidence of profound knowledge, a knowledge which could but be the outcome of patient and unwearied research. No better testimony of his many-sided genius can be adduced than that of Collins, who, on sending him a copy of some mathematical works, wrote : " Not that I think any man's works can add to your vast treasure of knowledge, but possibly may excite you to supply what they have omitted, or amend what they have but perfunctorily performed " *. Collins, it will be remembered, maintained an extensive correspondence with almost every other scholar of his day both at home and abroad, and there can be no doubt that his judgments are to be regarded as the embodiment of opinions then current. Halley, ever cautious in his commendations, addressed Wallis as " that judge .... to whom all others do and ought to subscribe " †. As we look back, after the lapse of so many years, we can now see Wallis, a titanic figure even in the midst of that brilliant group which made the latter half of the seventeenth century pre-eminent in the annals alike of science and of human progress.

For what might appear his comparative obscurity in the mind of the twentieth century reader two reasons may be suggested. Was it that the compass of subjects over which his genius ranged was too vast to allow of conspicuous success in any single one ? Or was it that, born into an age of intellectual giants, Wallis could

---

* Collins to Wallis, Jan. 2, 1665. (Rigaud : ii, p. 458–9.)
† Halley to Wallis, Nov. 13, 1686. (MacPike, E. F. : *Correspondence and Papers of Edmond Halley*, p. 70, 1932.)

not outstrip his contemporaries ? Certainly the period in which he lived was an epoch which will ever remain resplendent in the history of science. It was an age of varied intellectual activity. A sense of criticism was quickened, tempering the imagination, and frowning upon obedience to mere authority. A spirit of adventure seemed to be roused in the minds of men, urging them to explore the wide fields of knowledge. This reached its climax during the seventeenth century ; indisputably, no century starting from so little achieved so much as did this. On its very threshold Gilbert had enriched the world of learning with *De Magnete*, and within a few years of this auspicious opening the invention of the telescope had provided astronomy with a new and powerful stimulus. While Francis Bacon was philosophizing about the new experimental methods Galilei was putting them into practice, and already he had challenged the omniscience of Aristotle. In this country the tradition of experimental science was maintained by two men : Harvey, whose *De Motu Cordis* marked yet another milestone in the progress of science, and Boyle, whose new conception of the elements laid the corner-stone of modern chemistry. With the same restless zeal the frontiers of learning were being pushed forward in France by Descartes, Fermat and Pascal. During the latter half of the century names such as Huygens, Halley, Hooke, Wren remind us that the passion for enquiry had lost none of its earlier virility. Then came Newton, who gathered up the threads of the tangled skein and showed us something of its orderly pattern. In such a company did Wallis find himself. Little wonder is it, therefore, if his reputation was soon overshadowed. His genius, brilliant though it was, was obscured by the dazzling splendour of his great Cambridge successor.

Nevertheless it would be untrue to suggest that on that account his biography lacks colour. The life of one so richly endowed could hardly fail to excite more than a passing interest, albeit it was a life unmarked by those dramatic events which have shed a lustre round less brilliant men. His career was extraordinary in many ways, and not the least of these is the light it throws upon the temper of the age in which he lived. Although it was never granted to him to enrich the world of learning with some oustanding discovery, Wallis more than once pointed the way to what proved to be new and untraversed realms of thought; for it was by following the trail which had been blazed by Wallis that Newton was led to one of his most famous discoveries, the Binomial Theorem.

John Wallis was born on November 23rd, 1616, at Ashford, in East Kent, a county town of which his father was Rector. In 1622 his father died, and Wallis, who was the eldest son, was put to school at Ashford. Three years later, however, a severe pestilence

CHAPTER I

devastated the district, and Wallis was accordingly removed to Ley Green, in the parish of Tenterden. Here he came under the influence of one Mr. James Movat, a native of Scotland, by whom he was well grounded in the rudiments of grammar. Even at this early age he distinguished himself by that singular aptitude for learning which persisted with him till the closing years of his life. " It was always my affectation, even from a child ", he wrote to Doctor Smith, " in all pieces of Learning or Knowledge, not merely to learn by rote, which is soon forgotten, but to know the grounds or reasons of what I learn; to inform my Judgement as well as furnish my Memory, and thereby make a better Impression on both " *.

In 1630 Movat decided to close his school, that he might act as tutor to two of his pupils who were going abroad. He was naturally anxious that the young Wallis, who had already shown such promise, should accompany them, but this plan met with little encouragement from the boy's widowed mother. Accordingly Wallis was moved to Felsted, where he came under the care of Mr. Martin Holbech. From his letter to Doctor Smith we learn that Mr. Holbech told his new pupil that he came " the best grounded of any Scholar that he received from another School " †. During his first two years' stay at Felsted Wallis acquired a marked proficiency in Latin and Greek, " as wel qualified for the University as most that come thither " ‡. This we can readily believe, for his voluminous writings show him equally at home with Latin as with his native language. Hebrew also seems to have had an attraction for him, for he tells us that he rapidly acquired such a command over it as to be able to read the Bible in that tongue. These, together with the study of logic, seem to have constituted his principal occupations. Mathematics appears to have been virgin soil to him, and his initiation into that branch of learning did not take place until the Christmas of 1630, and then it was almost fortuitous. It is by no means uncommon in the history of science to find its votaries drawn almost imperceptibly to the road that has led to great achievements. Newton, according to Brewster, was first attracted to the pursuit of mathematics by nothing more than a desire to enquire into the truth of judicial astrology §; Wallis's interest in mathematics received its first impulse from circumstances no less commonplace. It is recorded that about this time a younger brother had been learning " to write and cipher and to cast account ". Moved by idle curiosity, Wallis

* Hearne, p. 144.
† *Ibid.*, p. 145.
‡ *Ibid.*, p. 146.
§ Brewster, p. 21.

enquired what all this meant, and was told that it consisted of "The Practical Parts of Common Arithmetick in Numeration, Addition, Substraction, Multiplication, Division, the Rule of Three (Direct and Inverse), the Rule of Fellowship (with and without Time), the Rule of False-Position, Rules of Practise, and Reduction of Coins, and some other little things "*. Wallis made himself master of this art in less than a fortnight, and this, he asserts, was his first insight into mathematics, and all the teaching he ever had. Assuredly it was a modest beginning for one who was to become, more than any other, the precursor of the mighty Newton. But so well did the subject suit his humour that on his return to Felsted he lost no opportunity of indulging in the pursuit of mathematical science. Such opportunities, however, came his way but rarely; there were, it seems, few to guide his early footsteps in this new territory. Mathematics at that period was scarcely regarded as worthy of the attention of the scholar; rather was it looked upon as part of the equipment of the mechanic, the merchant and the like. Even Francis Bacon, it will be remembered, in his elaborate scheme of learning hardly assigned to mathematics a position of primary rank. Scholasticism had relaxed its stranglehold, but in spite of a few zealous pioneers, whose names will presently be mentioned, its effects still lingered. The serious student was still expected to make proficiency in rhetoric, logic and the classics his first care, and it was to the study of these that Wallis's early activities were directed.

These, then, were the preoccupations of the young Wallis when, in 1632, he went up to Emmanuel College, Cambridge. Here he came under the direction of Mr. Anthony Burgess, "a pious, learned, and able Scholar, a good Disputant, a good Tutor, an eminent Preacher, a sound and orthodox Divine " †. The training he had received was such that he rapidly acquired the reputation of being an able disputant. "I soon became master of a *Syllogism*, as to it's true structure and the reason of it's consequences, however cryptically proposed, so as not easily to be imposed upon by Fallacies or false Syllogisms, when I was to answer or defend: and to manage an Argument with good Advantage, when I was to Argue or to Oppose, and to Distinguish ambiguous Words or Sentences as there was occasion " ‡. From logic he followed the usual course of study, which by then embraced ethics and metaphysics, but he did not hesitate to leave the narrow limits prescribed by tradition. He studied anatomy under Doctor Glisson, and he was the first

* Hearne, p. 147.
† *Ibid.*, p. 148.
‡ *Ibid.*, p. 149.

CHAPTER I

of that Professor's students to maintain in public disputation the doctrine of the circulation of the blood. Moreover, his latent talent for mathematical pursuits manifested itself in an eagerness to turn from the course of learning then in fashion to the study of astronomy.

Yet, although he could not forbear to indulge in the pursuit of mathematical learning whenever an opportunity presented itself, his early efforts seem to have been fitful. If we are to rely on his own testimony, this was because there was no one to direct him. More than once he laments the poverty of the mathematical training at Cambridge, and he observes that of the two hundred students in his college he could not name two whose mathematical learning surpassed his own, feeble though that was. In the whole University there were not many more. Wallis complains that the study of mathematics was prosecuted much less vigorously in Oxford and in Cambridge than it was in London, and he submits this as a reason for the mathematical sterility of his own college. But why he should maintain such a view is not easy to understand, for during this century the mathematical schools of the older Universities seem to have displayed marked originality and vigour. Certainly Cambridge was the real birthplace of Wallis's mathematical genius, as it was to be some thirty years later for Newton. Her teachers fostered his earliest studies, and it was within her walls that he acquired enough mathematics to be accounted the peer of Descartes, Pascal and Fermat. Not only that, but even if we except Newton, the list of mathematicians whom Cambridge honours as her *alumni* is no mean one. Briggs, Oughtred, Seth Ward, Horrocks, Pell, Barrow—all these, and many others, received their mathematical training at Cambridge. Nor did Oxford lag behind, and indeed, it would be illiberal to reproach a University whose mathematical strength is testified by names such as Wren, Neile, Halley, Rooke. Moreover it must not be forgotten that this period was marked by the establishment of important professorships in mathematics. At Oxford Sir Henry Savile had founded in 1619 the Savilian Professorships of Geometry and Astronomy, and less than half a century later the Lucasian Professorship was founded at Cambridge by Henry Lucas.

In the Hilary Term of 1636–37 Wallis graduated Bachelor of Arts, and four years later he was admitted to the Master's degree. Apparently his genius was now beginning to be recognised, for an effort was made about this time to elect him to a Fellowship of his college. The Statutes provided, however, that there should be only one Fellow from each county, and Kent was already represented by a Mr. Wellar. " Otherwise ", wrote Wallis, " I was well esteem'd and well-beloved in the College, and had certainly been chosen *Fellow*, if I had been in a capacity for it and loth they were that I should go away"*. Doctor Oldsworth, then Master of Emmanuel College, even went so far as to try to found a new Fellowship rather than risk losing one who showed such promise. But the project did not mature, and Wallis was eventually chosen Fellow of Queens' College, whither he removed soon afterwards.

In spite of his predilection for mathematics Wallis had always made Divinity his principal interest, and even as a boy he had looked forward to taking Holy Orders in the Church of England. From his very earliest years he had been given a strictly religious training, during which, as we gather from his letters, he was not only preserved from vicious courses and acquainted with religious exercises, but he was also instructed in the principles of religion and catechetical Divinity, the frequent reading of Scriptures and the diligent attendance on sermons. In 1640 he was ordained by Dr. Walter Curll, who held the See of Winchester, and soon afterwards he left his college on his appointment as chaplain to Sir Richard Darley at Buttercramb, in Yorkshire. A year later he left Yorkshire to act as private chaplain to Lady Vere, the widow of Lord Horatio Vere.

It was during the exercise of his duties with this family that an opportunity arose of exhibiting his skill in yet another art, namely, that of deciphering cryptic messages. The country was at this time in the turmoil caused by the Civil War. A chaplain to Sir William Waller chanced to show a letter written in cipher which had fallen into his hands after the taking of Chichester, December 29th, 1642. In an incredibly short time Wallis laid bare the message, and it is thought, and not without reason, that it was this fresh manifestation of his ability that brought him to the notice of the Parliamentary party, to which Wallis was an adherent ; in any case, he was presented, before the age of thirty, to the living of St. Gabriel's, Fenchurch Street. Here he seems to have acquired a certain controversial notoriety, which was to cling to him all through life, when he published *Truth Tried, or Animadversions on the Lord Brooke's Treatise on the Nature of Truth*.

In 1644 he was appointed, together with Adoniram Byfield, Secretary to the Assembly of Divines at Westminster, a post which he continued to occupy for as long as that Assembly held its sittings. From the conversations and the learned debates of so many grave and reverend Divines Wallis acknowledged that he derived much benefit. This Assembly had been convened by Parliament, who, having resolved upon the abolition of Episcopacy, looked to the Assembly to suggest a suitable alternative. So apt a pupil did

* Hearne, p. 151.

Wallis prove himself that within a few years we find him engaged in stoutly vindicating the Assembly from various charges that had been directed against it.

In the same year (1644) he married Susanna, daughter of John and Rachel Glyde, of Northiam in Northamptonshire, by whom he had (besides other children who died in infancy) a son and two daughters.

Wallis was now approaching his thirtieth year, but so far his youthful promise had not been fulfilled. Many causes conspired to bring about an almost complete relaxation of his scientific ardour. In the first place his ecclesiastical activities, which were daily increasing, must have absorbed much of his time and energy. Moreover, in those turbulent days opportunities of indulging in scientific pursuits rarely came his way. The Civil War, which had already demoralised the country, was fast approaching a critical stage. Not unnaturally, academic studies at both Oxford and Cambridge received a severe check. Fortunately, however, there arose a circumstance which led to a re-orientation of Wallis's activities, namely, the weekly meetings to which the Royal Society owes its inception. In his capacity as Secretary to the Assembly of Divines Wallis had been in intimate contact with several scholars who were eager to learn something of Natural Philosophy as well as of other branches of knowledge. Particularly were they attracted to the new Experimental Philosophy, the lofty conception of which had already been proclaimed by Francis Bacon. It is from the weekly meetings of these enthusiasts that can be traced the development of the Royal Society, and as Wallis's career pivoted round the activities of this illustrious institution to an extent hardly realized, an account of its origin and its early history will be necessary for a right understanding of the influence he exerted upon the scientific thought of his day. This will therefore mark a convenient point from which to look back upon the modest beginnings of that distinguished assembly of scientific inquirers—" the choicest witts in Christendome, and the finest parts " *—as Huygens afterwards dubbed them,—which became the dominant feature of the scientific work of the latter half of the seventeenth century.

* Huygens, vii, p. 11.

CHAPTER II.

The Beginnings of the Royal Society. Wallis's Part in its Early History. Appointment as Savilian Professor of Geometry.

The Royal Society arose out of the informal gatherings of the votaries of Experimental Science which took place about the middle of the seventeenth century. The new learning, at last freeing itself from the bondage of Aristotelianism, was finding its way into the Universities, and the number of scholars who were beginning to devote their attention to Natural Philosophy was fast increasing. One unmistakable sign of the passion for enquiry which was rapidly spreading over Europe is to be found in the establishment of societies of men who gathered together for the purpose of discussing the new learning and encouraging its development. England lagged behind the Continent in this respect, and it is to Italy that we must turn for the first organised scientific academy *. Not until 1645, by which time the Italian Society had been in existence for the greater part of a century, do we find any account of those meetings which eventually grew into the Royal Society. These meetings were doubtless inspired by that insatiable desire for knowledge which had been stimulated by the *Novum Organum*, but they also owe much to the institution founded under the will of Sir Thomas Gresham in 1579, whereby seven professors were employed to give lectures at Gresham College on successive days of the week, on Divinity, Astronomy, Geometry, Physic, Law, Rhetoric, Music. Wallis was keenly interested in the new venture, and of those others whose zeal prompted them to associate themselves with it probably the best known were Dr. Wilkins, afterwards Bishop of Chester, and Doctor Jonathan Goddard, Gresham Professor of Physic. Other distinguished names were those of Doctor George Ent, Doctor Glisson and Doctor Merrett, and Mr. Samuel Foster, then Professor of Astronomy at Gresham College. It was, however, a native of the Palatinate, a Mr. Theodore Haak, who was to be its *fons et origo*. It was he who gave the movement its first impulse by suggesting that these enthusiasts should

---

* The first Society established for the discussion of Physical Science was the *Academia Secretorum Naturæ* (Naples, c. 1560), under the Presidency of Della Porta. This was followed, in 1603, by the more famous *Accademia dei Lincei*.

CHAPTER II

meet together and form a society, not only for the promotion of those branches of learning in which they were specifically interested, but also for the purpose of examining "all systems, theories, principles, hypotheses, elements, histories, and experiments of things naturall, mathematicall, and mechanicall, invented, recorded, or practised, by any considerable author, ancient or modern" *. The pioneers were men of varied intellectual ability, and there was scarcely any avenue of learning which they did not explore. Even in those early days, therefore, the influence of such a society of enthusiasts in focussing scientific opinion, in making known the investigations of its members, and "in order to the compiling of a complete system of solid philosophy for explicating all phenomena produced by nature or art, and recording a rationall account of the causes of things" †, must have been a powerful stimulus to the development of science during that peculiarly difficult period.

Usually these meetings took place at Doctor Goddard's in Wood Street, because he kept an operator who could grind glasses for telescopes and microscopes. Sometimes the members met at some convenient place in Cheapside, sometimes at Gresham College. In 1648 some of the company migrated to Oxford, where, three years later, they assumed the name of the Philosophical Society of Oxford. The London section continued to meet at Gresham College until 1658, when the building was used as a barracks.

Some interesting observations regarding the enthusiasm of these men in their scientific researches are to be found among the papers of Robert Boyle, and there appears to be no doubt that the "Invisible College" which he so frequently eulogises was but another name for those gatherings to which we have referred. "The corner-stones", he wrote from London (Feb. 20, 1646/7), "of the invisible, (or as they term themselves) the *philosophical college*, do now and then honour me with their company" ‡. But a much more entertaining account of the beginnings of the Royal Society is given by Wallis himself in a curious tract entitled *A Defence of the Royal Society : An Answer to the Cavils of Doctor William Holder, published in 1678.* Doctor Holder had affirmed that "divers ingenious persons in *Oxford* used to meet at the lodgings of that excellent Person, and zealous promoter of Learning, the late Bishop of *Chester*, Doctor Wilkins, then Warden of *Wadham* .... where they diligently conferred about Experiments in Nature and Researches and indeed laid the first ground and foundation of the Royal Society" §. In answer to this Wallis wrote : "I take

* Statutes (Weld, i, p. 147).
† *Ibid.*
‡ Birch (1744), i, p. 20.
§ Preface by the publisher of the *Philosophical Transactions*, 1678.

its first ground and foundation to have been in London about the year 1645, (if not sooner) when the same Doctor Wilkins, (then Chaplain to the Prince Elector Palatine in London) .... with myself and some others met weekly, (sometimes at Doctor Goddard's lodgings, sometimes at the Mitre in Wood Street hard by), at a certain day and hour, under a certain penalty, and a weekly contribution for the Charge of experiments, with certain Rules agreed amongst us. Where (to avoid diversion to other discourses, and for some other reasons,) we barred all Discourses of Divinity, of State affairs and of News (other than what concerned the business of Philosophy) confining ourselves to Philosophical Inquiries and such as related thereunto, as Physick, Anatomy, Geometry, Astronomy, Navigation, Staticks, Mechanicks and Natural Experiments.

"These meetings we moved soon afterwards to the Bull Head in Cheapside, and in (Term time) to Gresham College, where we met weekly at Mr. Foster's lectures (then Astronomy Professor there) and after the lecture ended, repaired sometimes to Mr. Foster's lodging, sometimes to some other place not far distant, where we continued such enquiries, and our numbers increased.

"About the year 1648–49, some of our Company were moved to Oxford, first Doctor Wilkins, then I, and soon after, Doctor Goddard. Doctor Ward, Doctor Petty and many others of the most inquisitive persons met at Oxford at Doctor Petty's lodgings so long as Doctor Petty continued at Oxford, and for some time afterwards because of the convenience he had there (being the house of an apothecary) to view and make use of drugs, and other like matters as there was occasion.

"Our meetings there were very numerous and very considerable. For besides the business of persons studiously inquisitive, the Novelty of the Design made many to resort hither .... We did afterwards (Doctor Petty being gone to Ireland) and our Members grown less, meet at Doctor Wilkin's lodging in Wadham College. In the meantime our company at Gresham College being again much increased by the accession of divers eminent and noble persons, upon His Majesty's return we were (about the beginning of the year 1662) by His Majesty's grace and favour, incorporated by the name of the Royal Society."

The first Charter of Incorporation passed the Great Seal on July 15, 1662. This is therefore the date of the Royal Society. The Charter was read before the Society on August 13th of the same year, and on the 29th the President, the Council and the Fellows went to Whitehall and returned thanks to the King. The first Charter, however, did not give the Fellows all the privileges they desired, and representations being made, a second Charter,

CHAPTER II

supplying all the desired privileges, passed the Great Seal on April 22, 1663, and was read before the Society on the 13th of May following. It is by this that the Society has since been governed.

It is, however, not merely as a pioneer of the new movement that Wallis is deserving of credit. The early history of the Royal Society was not a uniform course of progress, and many of its members, only a small proportion of whom were, after all, men of science, were but lukewarm towards the new venture. Boyle's opinion of the founders, " men of so capacious and searching spirits, that school-philosophy is but the lowest region of their knowledge " *, must be accepted with caution. Evelyn, for example, twice refused the office of President, though " much importun'd ". The pioneers of the new learning were in a hopeless minority and were frequently assailed with vituperation from pulpit and from platform. Worse still, the Society had barely taken shape before it became a target for shafts of ridicule and derision which were launched by some of the ablest and best known men of the day. Even its Royal founder, as Pepys records (Feb. 1, 1663/4) " mightily laughed at Gresham College, for spending time only in weighing of ayre, and doing nothing else since they sat ". This was the golden age of wits and satirists, and the early pages of the Journal Books of the Society reveal a wealth of material for their jibes. Sometimes a note of singular bitterness is apparent : Steele, for example, wrote (*Tatler*, no. 236, Oct. 12, 1710) : " I have made some observations in this matter so long that when I meet with a young fellow that is an humble admirer of the sciences, but more dull than the rest of the company, I conclude him to be a Fellow of the Royal Society ". Swift, at greater length and more laboured sarcasm, caricatured the philosophers in his *Voyage to Laputa*, whilst Pope pilloried the small band of enthusiasts with defter touch. Even as late as 1669 this attitude towards the new-fledged Society persisted, for one of Wallis's letters, written in the summer of that year, tells us that on the occasion of the Dedication of the New Theatre Doctor Robert Smith, a University Orator, execrated the new philosophy in a long oration which consisted of "Satyrical invectives against Cromwell, Fanaticks, *the Royal Society and new Philosophy* " †. No wonder Bishop Sprat found it necessary to devote a whole volume of his history to a justification of the Society's aims. " I will now proceed ", wrote that distinguished historian of the Society, " to the weightiest, and most solemn part of my whole undertaking ; to make a defence of the *Royal Society*, and this new *Experimental learning*, in respect of

* Birch (1744), i, 20.
† Wallis to Oldenburg, July 16, 1669. (*Royal Society Guard Books*.)

the *Christian Faith* .... The *Experiments* of *Natural things* do neither darken our eies, nor deceive our minds " *. Perhaps more disastrous even than all this was the fact that troubles both at home and abroad rendered the early existence of the Society precarious ; disturbances of such magnitude could but react harmfully upon the progress of science, and during the last two or three decades signs were not wanting that academic studies were beginning to languish. It was during this critical period that Wallis devoted his whole energies to the task of stimulating the enthusiasm of the infant Society. The many letters which he despatched whilst at Oxford to Oldenburg, one of the Society's first secretaries, is evidence of his tireless zeal. The range of these letters is truly encyclopædic, and they show, as nothing else can, Wallis's devotion to the interests of the Society. It will be recalled that the Society's original plan had been to develop the technique of experimentation ; how faithful Wallis was to this empirical spirit appears on almost every page of these letters. Experiments which could not be conveniently carried out in London were performed at Oxford, where, it appears, greater facilities for such research existed, and to Wallis was assigned the important task of keeping the other members informed of the progress which was being made at Oxford. Here the new method of enquiry was leading its votaries along many paths, and as it was difficult, in the absence of scientific journals, for the scholar of those days to keep abreast of current developments, the dissemination of scientific news was a matter of first importance. From his letters we gather that Wallis zealously applied himself to almost every branch of learning. He made astronomical observations and submitted them to the Society through Oldenburg ; he elaborated his theory of the Flux and the Reflux of the Sea ; he described experiments on blood transfusion, and he sent Oldenburg his observations on the height of the barometer at different seasons, together with the highly interesting, and in some cases erudite speculations as to the reason of the wayward behaviour of the quicksilver in the tube. Perhaps more interesting than any of these were his many letters containing his observations on Gravity, a subject to which he had devoted profound attention. These, however, represent but a fraction of the problems he attacked, and from a review of his activities over this period, which is revealed by his voluminous correspondence †, the indisputable fact emerges that full justice has not been done to Wallis for his zeal in infusing so much vitality into the newly

* Sprat, pp. 345, 347.
† As well as his contributions to the *Philosophical Transactions*. See Appendix III.

formed Society. It must not be forgotten that the country was still in a chaotic state. Civil as well as religious discord was so rife as to present an almost insuperable barrier to the establishment of any permanent institution for the development of learning. The hope that the Restoration would settle men's minds and leave them free for philosophical pursuits had not been realized. The bulk of the English nation, whirled in the vortex of fierce religious and political dissensions, stood aloof from the momentous developments which were taking place in their midst, and there can be no doubt that had it not been for what Evelyn describes as " that learned Junto .... who more peculiarly applied themselves to the examination of the so long domineering methods and jargon of the Schools " *, the early history of the Society would have been a melancholy tale of decline. Of that small band of enthusiasts none was more energetic or more deserving of praise than was Wallis.

Now although Wallis's interests ranged over nearly every branch of human activity, it was in mathematics that he was to add imperishable lustre to his own name and to that of his country. Yet, as we have seen, during the first dozen years of his academic career Wallis had given scarcely any indication of his predilection for mathematics, and it was not until 1647 that his interest in this branch of science was revived. In that year a copy of Oughtred's *Clavis Mathematicæ* came into his hands. The *Clavis* was undoubtedly a distinct contribution to the mathematics of the period, though it hardly deserved the eulogies which Wallis lavished upon it. Writing to Collins in 1667 he spoke of it as a lasting book, and Oughtred himself as a classic author. " But for the goodness of the book in itself, it is that (I confess) which I look upon as a very good book, and which doth in as little room deliver as much of the fundamental and useful part of Geometry (as well as of Arithmetic and Algebra) as any book I know " †. Moreover the treatise seems to have made a lasting impression upon Wallis, for in the Preface to his *Algebra*, written nearly forty years later, when Wallis was an established mathematician, the encomiums are repeated. Whether Wallis was unnecessarily extravagant in his admiration is a question which will be discussed later ; the important fact is that it proved such a powerful stimulus to him that he promptly set about mastering it, and within a few weeks had so far succeeded as to discover for himself a new method of resolving cubic equations. This achievement, which he at once communicated to Mr. John Smith, Fellow of Queens' College, and

* Evelyn, p. 348.
† Wallis to Collins, Feb. 5, 1666/7 ; Rigaud, ii, 475.

Professor of Mathematics at Cambridge, gave unmistakable evidence of Wallis's mathematical strength. Towards the end of the following year the post of Savilian Professor of Geometry became vacant. This post had been held with distinction for many years by Doctor Peter Turner, but upon the outbreak of the Civil War in 1642 Turner had left Oxford to enlist under the Royalist banner with Sir John Byron. After the first skirmish at Edge Hill Turner was captured and removed to Northampton, where he was imprisoned. Meanwhile his possessions were ruthlessly seized by the Parliamentary Party ; in spite of this, however, his devotion to the cause of King Charles remained unshaken. But he paid dearly for his loyalty. By an Order of Parliament, dated November 9, 1648 *, he was ejected not only from his Fellowship but also from his Professorship. Wallis, in spite of the fact that he had incurred the lasting hostility of the Independents by signing the Remonstrance against the King's execution, was appointed to the vacant Professorship on June 14, 1649 †.

This appointment was most acceptable to him, for it gave him an opportunity of exercising his mathematical talents which had for so long remained dormant. Mathematics, he said, which had hitherto been a pleasing diversion, now became his serious study. He thereupon left his Church (St. Martin's, Ironmonger Lane) and went up to Exeter College, Oxford; and although his interest in his ecclesiastical duties was never allowed to wane, he applied himself with conspicuous zeal to the duties of his new office. He soon became President of the Oxford Philosophical Society, which began to hold its meetings about the end of the year 1651, and in that capacity he was enabled to maintain that intimate contact with the parent Society in London which was to be of such inestimable benefit to both. Indeed, a most cursory glance at the early Minute Books of both Societies reveals how numerous were the scientific contributions from the Oxford members, and what an active share Wallis took in maintaining the closest union between the two.

* Ward, p. 133.
† Wood, ii, Col. 753.

CHAPTER III.

The Treatise on the Conic Sections.

WALLIS's appointment to the post of Savilian Professor of Geometry marks the beginning of a period of intense mathematical activity, which lasted almost without interruption till the closing years of his life. About this time the subject of Indivisibles was very much in the air. Wallis's interest in this newly awakened branch of mathematics received its first impulse from a chance perusal of the works of Torricelli, in which Cavalieri's methods were freely employed. The subject interested him profoundly, for underlying it he thought he saw a means whereby the quadrature of the circle might be effected. For more than three years he devoted all his energies to an attempt at an elucidation of this centuries old problem. The outcome of his patient investigation was the *Arithmetica Infinitorum* *, which was published in 1656, and which at once placed its author in the foremost rank of mathematicians of even that illustrious age. This important treatise will be discussed later; it will not, however, be out of place to mention at this point its most salient features.

The *Arithmetica Infinitorum* relates primarily to the quadrature of curves by the *method of indivisibles* which had been elaborated by Cavalieri, and to which notable contributions had subsequently been made by Fermat, Descartes and Roberval. But where Cavalieri was limited to a geometrical exposition Wallis had to his hand the powerful tool which had been forged by Descartes nearly twenty years earlier. This was the Analytical Geometry, the principles of which had already been expounded by the great French geometer in 1637, and which had been rendered familiar throughout Europe by van Schooten. The method is substantially that of the Integral Calculus, and by the method of summation which he developed Wallis was enabled to perform operations which, if we substitute the modern terminology for that employed in the

---

* The full title is *Arithmetica Infinitorum sive Nova Methodus Inquirendi in Curvilineorum Quadraturam, aliaque difficiliora Matheseos Problemata*. (The Arithmetic of Infinities, or a New Method of studying the Quadrature of Curves, and other more Difficult Mathematical Problems. Oxon, 1656. (The date of its Dedication was July 19, 1655.)

pages of the *Arithmetica Infinitorum*, would be expressed by the relations

$$\int_0^1 (1-x^2)^0 \, dx = 1,$$
$$\int_0^1 (1-x^2)^1 \, dx = 2/3,$$
$$\int_0^1 (1-x^2)^2 \, dx = 8/15,$$
$$\int_0^1 (1-x^2)^3 \, dx = 48/105.$$

Now the area of the quadrant of a circle of unit radius is expressed by the integral $\int_0^1 (1-x^2)^{\frac{1}{2}} dx$, and thus the ratio of the area of the square on the radius to that of the circular quadrant is expressed by the relation

$$4 : \pi = 1 : \int_0^1 (1-x^2)^{\frac{1}{2}} dx.$$

This integral can be comprehended in the general form

$$\int_0^1 (1-x^{\frac{1}{p}})^n \, dx,$$

and if a solution of this integral could be obtained the problem of the quadrature of the circle would be solved.

To evaluate this integral Wallis advanced beyond his predecessors by making extended use of the "law of continuity", upon which he placed the fullest reliance, even extending it so as to include fractional and negative indices. By giving p and n a succession of values he obtained the corresponding values of this integral, and from a consideration of these he tried to discover what would be its value if p and n were each equal to $\frac{1}{2}$, as the quadrature of the circle demanded. By a complicated chain of reasoning he eventually arrived at the relation

$$\frac{4}{\pi} = \frac{1}{\int_0^1 (1-x^2)^{\frac{1}{2}} dx} = \frac{3 \cdot 3 \cdot 5 \cdot 5 \cdot 7 \cdot 7 \ldots}{2 \cdot 4 \cdot 4 \cdot 6 \cdot 6 \cdot 8 \ldots},$$

in which both numerator and denominator are continued products. To the *Arithmetica Infinitorum* we owe the word *Interpolation*, and also the name *Hypergeometric Series*, by which Wallis understood the series whose terms are

$$1, \frac{1 \cdot 3}{2}, \frac{1 \cdot 3 \cdot 5}{2 \cdot 4}, \frac{1 \cdot 3 \cdot 5 \cdot 7}{2 \cdot 4 \cdot 6},$$

and whose general term is

$$\frac{1 \cdot 3 \cdot 5 \ldots \ldots (2n+1)}{2 \cdot 4 \ldots \ldots 2n}.$$

CHAPTER III

Furthermore, Wallis obtained the relation which to-day would be written

$$ds = \sqrt{1+\left(\frac{dy}{dx}\right)^2}\,dx$$

for the length of an element of a curve, and thus he indicated a way in which the problem of rectification might be reduced to that of quadrature.

To the *Arithmetica Infinitorum* Wallis prefaced his Treatise on Conic Sections (*De Sectionibus Conicis, Nova Methodo Expositis, Tractatus, Oxon,* 1655). This work, which indisputably merits a far greater measure of recognition than has yet been accorded to it, presents an ancient subject in a new setting. " What new matter this little book brings forth ", he declared in the Dedication to Seth Ward and Laurence Rooke, " appears at once from the title which it bears. For it deals with the Conic Sections (an old subject) in a new fashion " *.

The Conics had been known and their properties had been investigated by the mathematicians of antiquity, all of whom, however, had regarded these figures purely as plane sections of a cone. The more ancient mathematicians, as, for example, Menæchmus and his School (4th Century B.C.), had admitted only the right cone into their geometry, and they considered the section of it to be made by a plane perpendicular to a generator. Now since the vertical angle of a right circular cone may be right, acute or obtuse, the same method of cutting, namely by a plane perpendicular to a generator, might produce all three Conic Sections. The parabola was called the section of a right-angled cone, the ellipse the section of an acute-angled cone, and the hyperbola the section of an obtuse-angled cone. Apollonius observed, however, that these three sections might be obtained in every cone, be it oblique or right, and that they depended upon the different inclinations of the plane of section to the cone itself.

But the methods of Apollonius were very obscure, and his expositions long and involved. Wallis therefore essayed to treat these figures analytically as plane curves, " without the embranglings of the cone ". In the Dedication he explained why he had been led to introduce his new methods. The ancients, he complained more than once, had been more anxious to be admired than understood, and Apollonius, though no one would deny him the title *Magnus Geometer*, was no exception. On account, therefore,

* " Quid novi ferat (Clarissimi Viri) Libellus iste, ab ipso statim quem præfert titulo innotescit. Nempe Sectiones Conicas (rem veterem) nova methodo tradit." Dedication.

of inherent difficulties (which later geometers had done nothing to remove) this important branch of mathematics had been allowed to fall into disuse. (*Nimis neglecta fuit, tanquam insuperabilis difficultatis plena.*) No better evidence of this neglect could be adduced, asserted Wallis, than the fact that of the eight books which Apollonius had contributed to the subject only four had survived. Wallis felt that he could treat the Elements of the Conics in a manner no more intricate than the Elements of Euclid. To effect this he dispensed with the cone altogether, and regarded the Conics merely as plane curves (" curvas in plano descriptas "). For, he maintained, it is no more necessary to regard the parabola, the ellipse, the hyperbola solely as sections of a cone, than it is to regard the circle as no more than the section of a cone by a plane parallel to the base, or even the triangle as the section by a plane through the vertex *. Not only that, but to Wallis the Cone was not merely a solid of revolution, as it was to the earlier geometers, to him it was the figure made up of an infinite number of circles or thin cylinders of regularly diminishing diameters.

In this treatise Wallis first gave an illustration of his new Method of Indivisibles, and in its pages he applied his infinite numerical progressions to the geometrical results obtained by his predecessors. It had frequently been alleged that the methods adopted by Cavalieri were not altogether above reproach, inasmuch as he had been accustomed to regard a line as made up of an infinite number of points, a surface of an infinite number of lines, and so on. Wallis, however, attempted to free the method of this criticism by considering a plane to consist of an infinite number of parallelograms each of indefinitely small altitude. This new conception is introduced on the threshold of the treatise. " To begin with ", he says in his first proposition, " I suppose any plane (following the *Geometry of Indivisibles* of Cavalieri) to be made up of an infinite number of parallel lines, or as I would prefer, of an infinite number of parallelograms of the same altitude ; (let the altitude of each one of these be an infinitely small part, $\frac{1}{\infty}$ of the whole altitude, and let the symbol ∞ denote Infinity) and the altitude of all to

---

* " Non enim est *Parabolæ* magis essentiale, ut fiat *Sectione Coni plano lateri parallelo* ; quam *Circulo*, ut fiat *Sectione Coni plano basi parallelo* ; aut *Triangulo*, ut fiat Sectione Coni per verticem." (It is no more necessary to regard a Parabola as the section of a cone by a plane parallel to a side than it is to regard a circle as the section of a cone made by a plane parallel to the base, or a Triangle by a section through the vertex. (Prop. 21.)

CHAPTER III

make up the altitude of the figure " *. Now, although the substitution of infinitely small parallelograms for the parallel lines of Cavalieri had already been hit upon by others, notably Pascal and Roberval, it nevertheless remained for Wallis to introduce a conception which was to prove of inestimable value to mathematicians later. In showing a preference for his " line " over that of Cavalieri he observed that his line was supposed to be *dilatable*, or to have so much *thickness* as that by an infinite multiplication to be capable of acquiring a certain altitude equal to that of the figure in which it was inscribed †. Here the word *dilatable* (*dilatabilis*) conveys an idea very different from the increase made by the apposition of parts, and clearly shows that he conceived these little parallelograms increasing by dilation until they filled the whole space. This, of course, is nothing less than the notion of the generation of quantities which is the basis of the Method of Fluxions, and it is interesting to note a similar view in a letter to Leibniz, in which Wallis observes that his Method of Tangents might justly be called *Methodus de Magnitudine Evanescente*.

In the Conic Sections, as in the *Arithmetica Infinitorum*, Wallis showed himself one of the first mathematicians to realize the meaning of the terms *infinity* and *infinitely small*. Though it would admittedly be an exaggeration to declare that he had clearly grasped the modern conception of infinity regarded as a limit, nevertheless in his frequently repeated *quavis assignabile minor* he first created the conception of a limit as something *becoming*, and this idea was highly significant in itself even if Wallis had been content to stop there. He speaks of the asymptotes of the hyperbola continually approaching the curve until the distance between them becomes less than any assignable quantity (*ut distantia earum tandem evadat quavis assignata minor*—Prop. xxxix). Unfortunately, however, he falls into the not uncommon error of regarding infinitely small and zero as synonymous expressions, *i. e.*, as though $a-a$ were the same thing as $\frac{1}{\infty}$. For example, on the threshold of the treatise

---

* " Suppono, in limine (juxta Bonaventuræ Cavalierii *Geometriam Indivisibilium*) Planum quodlibet quasi ex infinitis lineis parallelis conflari ; Vel potius (quod ego mallem) ex infinitis Parallelogrammis æque altis ; quorum quidem singulorum altitudo sit totius altitudinis $1/\infty$, sive aliquota pars infinite parva ; (esto enim ∞ nota numeri infiniti ;) adeoque omnium simul altitudo æqualis altitudini figuræ." (Prop. 1.)

† " In hoc saltem differunt, [mea methodus et Cavalierii] quod linea hæc supponitur dilatabilis esse, sive tantillam saltem spissitudinem habere ut infinita multiplicatione certam tandem altitudinem sive latitudinem possit acquirere, tantam nempe quanta est figuræ altitudo." (Prop. 1.)

we encounter the statement: "The altitude is supposed to be infinitely small, that is, no altitude, for a quantity infinitely small is no quantity, scarcely differing from a line "*. Similarly, his use of the term Infinity was not always above reproach. "If the first term of a series of Primana, *i. e.*, of the series

$$0+1+2+3+$$

be zero, the first term of the reciprocal series $\dfrac{1}{0}$ will be infinite, since in division if the divisor be zero the quotient will be infinite" †. In the *Arithmetica Infinitorum* he refers frequently to areas *greater than infinity* (*plusquam infinitas*), and even in the treatise *De Motu*, written fifteen years later, the expression recurs. Moreover, his "proof" of the formula for the area of a triangle can hardly be considered more rigorous than Cavalieri's, for, having divided the altitude A into an infinitely large number of parts, and having drawn parallel to the base B an infinite number of lines whose total length is $\dfrac{\infty \, B}{2}$, he writes

$$\text{Area } \frac{1A}{\infty} \times \frac{\infty \, B}{2} = \frac{1AB}{2}. \quad \text{[Fig. 1.]}$$

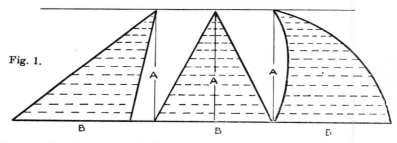

Fig. 1.

Frequently he treats infinity as though the ordinary rules of arithmetic could be applied to it. Thus he says: "For when an infinitely small quantity be multiplied by infinity, there emerges a quantity of some magnitude, for $\dfrac{1}{\infty} \times \infty = 1$, and $\dfrac{1A}{\infty} \times \infty = A$ " ‡.

---

* " Altitudo supponitur infinite parva, hoc est nulla, (nam quantitas infinite parva perinde est atque non quanta) vix aliud est quam linea." (Prop. 1.)

† " Cum enim primus terminus in Serie Primanorum sit 0; primus terminus in serie reciproca erit ∞ vel infinitus (sicut in divisione, si diviso sit 0, quotiens erit infinitus." (Prop. xci, Cor.)

‡ *Arith. Infin.* Prop. clxxxii, Scholium.

CHAPTER III

But this is perhaps understandable. For many years to come the greatest confusion regarding these terms persisted, and even in the next century they continued to be used in what appears to us as an amazingly reckless fashion.

Apart from such paradoxes much of the work is on distinctly modern lines. By enclosing the figure whose area he seeks between two figures, one inscribing it and the other circumscribing it, and by showing that as the number of sides increases the difference between them becomes an infinitely small part of the whole (*ut differentia tandem quavis assignata quantitate minor sit*), there was little that was new. But when he added *Hic quidem sive excessus sive defectus, quamdiu de finita series agitur omnino animadvertendus est, ubi autem de serie infinita agitur tuto poterit negligi* (p 152), *i. e.*, this excess or deficit must be taken into consideration so long as we are dealing with finite series, but it becomes negligible when the series is infinite, we are approaching, in a crude and imperfect form, perhaps, the principles of the Integral Calculus. Thirty years later he wrote : " We may observe a great difference between the proportion of *Infinite* to *Finite*, and, of *Finite* to *Nothing*. For $\frac{1}{\infty}$, that which is a part infinitely small may, by infinite Multiplication equal the whole. But $\frac{0}{1}$, that which is Nothing, can by no Multiplication become equal to Something " *.

To return to the *Conic Sections*. The treatise is divided into two parts. In the first part Wallis recalls the fundamental propositions relating to the Conics which had been taught by Apollonius. In the second part he shows how all these properties could have been determined without reference to the cone. " In what follows ", he declared at the beginning of the second part, " we may study these (either lines or figures) according to their inherent properties, without considering, as is usual, how they were generated either by sections of a cone, or in any other way " †.

His first task, therefore, is to define his curves anew, and then to show that the curves thus defined are the same as those obtained by the sections of a cone. Thus for the parabola : " I define the parabola ", he says, " as the curve described in a plane, (or the plane figure bounded by the curve OAO) whose intercepts are

---

* *Defense of the Angle of Contact*, p. 99.

† " Licebit in sequentibus illas (sive lineas sive figuras) secundum absolutam quam in se habent naturam contemplari, sine respectu ad genesin habito, quæ vel a Sectione Coni, vel etiam aliunde contingere possit."

as the square of the ordinates " *. That this curve is identical with the parabola of Apollonius is indicated in the same proposition †, and this is followed by an elaborate proof. Then follows an investigation, much in the style adopted in a modern treatise on the Conic Sections, of the properties of the curve, and these are summarized in the next proposition. Throughout the investigation he makes free and unrestricted use of the principles of analysis which had been developed by Descartes—a fact which many French writers, notably Chasles, do not fail to observe ‡.

The investigation of the properties of the ellipse, which begins at Proposition xxvi, is even more comprehensive. To define the curve he uses the property already known to Apollonius, namely, that if A and A' be two fixed points, and P a point which moves so that PM² (M is the foot of the perpendicular on to AA'), bears a constant ratio to AM . MA', then the curve traced out by P is an

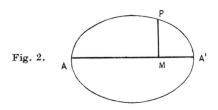

Fig. 2.

ellipse (fig. 2). Referring the curve to axes through the vertex, Wallis expresses this relation by means of an equation :

$$e^2 = ld - \frac{l}{t}d^2,$$

in which e is the length of the " ordinatim-applicata " (i. e., the ordinate) of the point, d its distance from the vertex (abscissa),

---

* " Parabolam igitur appello eam (sive lineam curvam in plano descriptam, sive figuram planam, ejusmodi curva terminatam, puta OAO) cujus ordinatim-applicatarum quadrata sunt interceptis-diametris proportionalia." (Prop. xxi.) The *ordinatim-applicata* to which Wallis so frequently refers is defined as the " half of the parallel right lines, (i. e. DO, PO,) terminated by each side of the curve, and bisected by the diameter ADP ".

† " *Parabolam* autem a *nobis definitam eandem esse cum Parabola Apollonii*, et aliorum ; talem nempe, qualis sectione Coni effici possit "— the parabola defined by me is the same as that of Apollonius and others, i. e., that which can be obtained by the section of a cone.

‡ *Chasles* ; Aperçu Historique. " Wallis écrivit, le premier, un *Traité analytique des sections coniques*, suivant les doctrines de *La Géométrie* de Descartes " (p. 101).

CHAPTER III

and l its latus rectum *. Hence the curve might also have been defined as the locus of a point which moves so that the square of its ordinates is proportional to the difference between two rectangles. Using this equation, Wallis is now enabled to investigate all the properties of the curve in a singularly beautiful and concise fashion, which was destined eventually to lead to the abandonment of the methods of the ancient geometers. The same treatment is employed in investigating the properties of the hyperbola, and the similarities between the properties of that curve and of the ellipse are clearly indicated.

The advantage of Wallis's mode of treatment is clear. " Once the fundamentals have been delivered ", he had claimed in the Dedication, " the remainder will give no difficulty ", and the boast was not extravagant. Moreover, his criticism of the methods of Apollonius—" which nevertheless still remained harsh and difficult, so that few would venture to approach them " †—was substantially accurate, although it must be admitted that Wallis was scarcely fair to some of his immediate predecessors. He seems to have forgotten the contributions of Claude Mydorge, the friend of Descartes, who, whilst still adhering to the ancient methods, had in his treatise *De Sectionibus Conicis* (1631) succeeded in introducing a marked explicitness into many of the proofs of Apollonius. Nevertheless this important branch of mathematics derived from Wallis's masterly exposition a benefit which cannot be easily estimated. Wallis himself was under no illusions as to the value of his treatise. Thirty years later he wrote : " How convenient it is thus to deliver the Elements of Conicks may be easily discerned by any who shall please to compare these with those formerly delivered by others. Yet I do not know that any before me had attempted it " ‡.

There is yet another feature of Wallis's treatment which commands attention. The seventeenth century had witnessed more than any other the development of the symbolic language of mathematics. This is illustrated in the writings of Hérigone and of Descartes in France, of Rahn in Switzerland, and of Harriot and Oughtred in England. The pioneers of this new movement found an enthusiastic disciple in John Wallis, for in this treatise he employed the new symbolism so freely as to bring him into open

---

* This is the modern equation to the ellipse referred to the vertex as origin, i. e., $y^2 = kx(a-x)$. Putting $ka = p$ we get $y^2 = px - \frac{p}{a}x^2$ (p is the latus rectum).

† " Aspera tamen adhuc mansit et difficilis, ut, qui illam ausus attingere, sit e paucis unus." Dedication.

‡ Algebra, p. 292.

conflict with its opponents, many of whom felt that compactness such as this might easily defeat its own purpose. In the van of his hostile critics stood his inveterate foeman, Thomas Hobbes. "And as for your Conic Sections", he wrote, "it is so covered over with the scab of symbols, that I had not patience to examine whether it were well or ill demonstrated"*. Again he wrote: "Symbols, though they shorten the writing, yet they do not make the reader understand it sooner than if it were written in words" †. Hobbes could hardly have made a more ill-judged remark, for the notation which Wallis adopted in the pages of the *Conic Sections* was such as to give a conciseness, and that without sacrifice of clarity, as was achieved by no other writer of the century. "In my choice of symbols", wrote Wallis, with refreshing modesty, "I have followed in part Oughtred, in part, Descartes". But the assertion was altogether too unpretentious. Turning aside from the excesses of Oughtred and the geometrical symbolism of earlier writers, he consistently used the Index notation, and by writing a^2 instead of aa he improved even upon Descartes. Further, it was in this treatise that Wallis gave to mathematics symbols which have since retained their currency. Chief of these is the symbol ∞ for infinity, whilst hardly less important is the symbol, first used by Wallis, $\geqslant$ for "equal to, or greater than". ("$DT \geqslant DO$, hoc est, DT æquales vel major quam DO").

The *Conic Sections* does not seem to have met with the welcome it deserved, and many years were to elapse before the methods introduced by Wallis were pursued as they should have been. A belated appreciation appeared in the *Philosophical Transactions* for 1695 (p. 73), where the work is described as "A Treatise of *Conick Sections*, in a new and Easie Method; considered as plain Figures, exempted out of the Cone. Whereby the Doctrine of the Conicks is made much more Intelligible than formerly, when it was esteemed so perplex and intricate as to deter many from meddling with it". A year or two after its publication Borelli, who occupied the Chair of Mathematics at Pisa, made a praiseworthy effort to render familiar the properties of the Conics in his *Apollonii Elementa Conica* (Pisa, 1658), but though his treatise was written with marked brevity and perspicuity it still adhered to the ancient methods. The same is to be said of Viviani's *Geometrica Divinatio in Quintum Librum Conicorum Apollonii Pergæi* (1659), which was an attempt to supply the deficiencies in the known works of Apollonius. Jan de Witt alone seems to have drawn inspiration from Wallis's treatise, for in the second part of his treatise *Elementa Curvarum Linearum* (1658) he employed the methods advocated by Wallis.

* *Molesworth*, vii, p. 316.
† *Ibid.*, vii, p. 248.

CHAPTER III

It is greatly to be regretted that the *Conic Sections* of Wallis was not accorded a warmer welcome. For, apart from the impulse it should have given to the study of these important curves, the treatise claims attention on other grounds. First, its investigations would have popularized the Cartesian notation even more than *La Géométrie*, with its deliberate obscurities, itself could have done. Secondly, by the clearness of its exposition it gave an impulse to a movement which was long overdue. Finally, the methods adopted by Wallis, in virtue of their generality, were not restricted merely to curves which could be formed by the sections of a cone. Wallis had indisputably opened up the way to an investigation of the properties of curves other than those already discussed. This he illustrates by an examination of the cubical parabola ($p^3 = l^2 d$) and even the biquadratic parabola ($p^4 = l^3 d$), the properties of which were easily investigated by the methods he had already explained.

If there is an explanation of the neglect with which this treatise was met it is probably to be found in the fact that it was overshadowed by the *Arithmetica Infinitorum*. Not for many years had the science of mathematics been enriched as it was by that treatise. We pass therefore at once to a consideration of the chief features of that monumental work, a work with which the name of John Wallis remains inseparably associated.

CHAPTER IV.

THE *ARITHMETICA INFINITORUM*.

BY the beginning of the seventeenth century the subject of Infinitesimals had already become one of profound study. Traces of an investigation into the subject are to be recognized among the writings of the ancients, and their ingenious attempts to determine the areas under curves mark the first step in its elucidation. But the difficulty of explaining the paradoxes of Zeno the Eleatic led the Greeks to look with some suspicion upon the notion of Infinitesimals, and later geometers, notably Eudoxus and Archimedes, and, after them, Pappus of Alexandria, sought to avoid their use by recourse to the somewhat cumbrous Method of Exhaustions.

Kepler, however, in his *Stereometria* (1615) took up the doctrine of Infinitesimals, and in this treatise he opened up a vast field of speculation into their nature. In determining the volumes of certain vessels bounded by curved surfaces he avoided the tedious method of exhaustions, and he fell back upon the doctrine that an area or a volume is made up of an infinite number of parts. Thus he conceived a circle to be formed of an infinite number of triangles having a common vertex at the centre of the circle, with their infinitely small bases in the circumference, and in this way he rendered familiar the ideas of infinitely great and infinitely small. Moreover, by applying the doctrine of continuity to these infinitesimals he may be said to have laid the foundations of the Integral Calculus. But what was probably more important still was the fact that his work attracted the notice of another ingenious mathematician, Bonaventura Cavalieri of Bologna. Cavalieri published his *Geometria Indivisibilibus* * in 1635. In this he considered a line to be made up of an infinite number of points, a surface of an infinite number of lines, and a solid of an infinite number of surfaces, and he laid it down as an axiom that " any magnitude may be divided into an infinite number of small parts which can be made to bear any required ratio (*e. g.*, equality) one to another ". This was a distinct advance. Roberval (1602–1675), and later Pascal (1623–1662), attempted to rid this doctrine of its obvious shortcomings by considering a line to be made up, not of points, but of indefinitely short lines. Similarly a surface was conceived

* *Geometria Indivisibilibus Continuorum Nova Quadam Ratione Promota*, Bononiæ, 1635.

to be made up of infinitely narrow parallelograms, and a solid of indefinitely thin solids. Pursuing this line of thought, Roberval and also Torricelli showed the area of the cycloid to be three times that of its generating circle. Fermat (1601–1665) carried the study of Infinitesimals still further, and on the principles of Cavalieri calculated the areas of several figures, mainly parabolas and hyperbolas. He also evolved a method of finding the maximum or minimum value of a function by a method which is substantially that of equating its first differential coefficient to zero.

Such was the state of this branch of mathematical science when Wallis directed to it the vigour of his mind. The result of his preoccupations is contained in the *Arithmetica Infinitorum*, which appeared in 1656, and which constitutes a landmark in the history of mathematics. The work was dedicated to Oughtred (*non tam Virum magnum quam magnum Mathematicum quærendum putavi, cui illud inscriberem*)*, and its publication at once placed Wallis in the front rank of mathematicians of even that illustrious age. And well indeed it might, for not only did it abound in fresh discoveries and new points of view, which to later mathematicians were to prove so stimulating, its whole contents clearly showed that mathematics had now reached a critical stage in its development, a development which was destined to make it before the end of the century the most potent instrument of physical research.

The Dedication is of very great interest, for in it Wallis reviews the work of his predecessors in this field. In particular, he explains how the methods of Cavalieri had attracted his notice as early as 1650, especially as he thought he saw underlying them a means of attacking the formidable problem of the quadrature of the circle. For three years he directed his whole energies to a solution of the problem, and although he submitted his difficulties to the leading mathematicians at Oxford, notably Seth Ward, then Savilian Professor of Astronomy, and to Christopher Wren, then Fellow of All Souls College, the elucidation of the problem continued to evade his grasp.

But in the *Arithmetica Infinitorum* Wallis came very near solving the riddle which had defied the most able mathematicians of even that century. The quadrature of many curves had been effected long before Wallis began his investigation. In no sense, therefore, can it be claimed that Wallis was a pioneer in this branch of mathematics. Cavalieri's methods, ill-founded though they were, had enabled him to derive a theorem which was equivalent

---

* " I thought that I should look not only for a great man, but rather a great mathematician to whom I should inscribe my work."

to finding the area under the curve $y=x^n$, and many other geometers—Fermat, Roberval, Torricelli—had shown, more or less independently of one another, that in this family of curves the area under the curve bore to that of the rectangle of the same base and altitude the ratio $\dfrac{1}{n+1}$. So long as the power was a whole number the methods established were sufficient; the difficulty arose when that power was either negative or fractional, as it was, for example, in the hyperbolic curves. Wallis's sagacity led him unerringly to perceive that the relation persisted for all values of n, positive, fractional, and even negative. By extending the "law of continuity" he was led to observe that the reciprocal of a power could be regarded as a power with a negative exponent, and by his *Method of Induction*, which he handled so skilfully, he was able to generalize, under one comprehensive law, the work of his predecessors. This done he turned to the quadrature of the circle, and although it was left for Newton to complete the investigation which Wallis had so brilliantly initiated, this does not in the least detract from the magnificence of his achievement, for the problem which he had set himself was no less difficult than the one which was part of his legacy to Newton.

If, however, we are to appraise in a just perspective the precise nature of Wallis's contribution, it will be necessary to follow his investigation step by step through the pages of this monumental treatise. For a statement of the main problem, and the steps which led to it, we turn again to the Dedication. By pursuing Cavalieri's method of Indivisibles, Wallis was already familiar with the following results:

(1) The aggregate of the circles which make up the cone is to the aggregate of the circles which make up the circumscribing cylinder (*i.e.*, the volume of the cone is to the volume of the circumscribing cylinder) as 1 is to 3. Since the diameters of the circles which make up the cone increase arithmetically (as 1, 2, 3, etc.), the circles themselves increase in the duplicate ratio of these numbers, *i.e.*, as 1, 4, 9, etc. Hence the above result is equivalent to the statement:

$$\frac{0^2+1^2+2^2+3^2+4^2+\cdot+\cdot+\cdot+n^2}{n^2+n^2+n^2+n^2+n^2+\cdot+\cdot+\cdot+n^2}=\frac{1}{3}.$$

(2) If we consider the triangle formed by taking a section of the cone through its axis, and the rectangle formed by taking a similar section through the circumscribing cylinder, it is known that the aggregate of the lines which make up the former is to the aggregate of those which make up the latter as 1 is to 2. But

CHAPTER IV

as the former of these increase arithmetically, this result may be written :

$$\frac{0+1+2+3+.+.+n}{n+n+n+n+.+.+n} = \frac{1}{2}.$$

(3) In the parabola (*i. e.*, the section of the conoid paraboloid) the aggregate of the lines making up the parabola is to the aggregate of the lines making up the circumscribing rectangle as 2 is to 3. But as the distances of these lines from the vertex increase arithmetically, *i. e.*, as 1, 2, 3, . . ., the lines themselves increase as the subduplicate of these numbers. That is, the above result is equivalent to the statement :

$$\frac{\sqrt{0}+\sqrt{1}+\sqrt{2}+\sqrt{3}+\sqrt{4}+\ldots+\sqrt{n}}{\sqrt{n}+\sqrt{n}+\sqrt{n}+\sqrt{n}+\sqrt{n}+\ldots+\sqrt{n}} = \frac{2}{3}.$$

(4) If we consider the sphere and its circumscribing cylinder it is known that the aggregate of the circles which make up the former is to that of the circles which make up the latter as 2 is to 3. But the circles which make up the sphere form the series (starting from the greatest) :

$$(R^2-0^2),\ (R^2-1^2),\ (R^2-2^2),\ (R^2-3^2),\ \ldots\ (R^2-R^2),$$

so that the above result is equivalent to the statement :

$$\frac{(R^2-0^2)+(R^2-1^2)+(R^2-2^2)+(R^2-3^2)+\ldots+(R^2-R^2)}{R^2+R^2+R^2+R^2+\ldots+R^2} = \frac{2}{3}.$$

Now, says Wallis, in the quadrant of the circle we have to consider the aggregate of all the lines from the radius down to 0. Since each of these lines is a mean proportional between the segments of the diameter, they form a definite series, starting from the greatest, namely :

$$\sqrt{R^2-0^2}+\sqrt{R^2-1^2}+\sqrt{R^2-2^2}+\sqrt{R^2-3^2}+\ldots+\sqrt{R^2-R^2}.$$

Therefore, if we can find the value of the ratio

$$\frac{\sqrt{R^2-0^2}+\sqrt{R^2-1^2}+\sqrt{R^2-2^2}+\sqrt{R^2-3^2}+\ .\ +\sqrt{R^2-R^2}}{R+R+R+R+.+R},$$

we have got the ratio of the area of the quadrant of the circle to the area of the square on the radius, *i. e.*, $\frac{\pi}{4}$. " Si inde cognosci posset quam rationem habeant omnes eorum diametri ad diametros horum inventum foret quod quæritur: nempe Circulum illæ, hæ Quadratum diametri constituunt " *.

This, then, is the problem, and the investigation into it opens with the proposition: "If a series of quantities be proposed, starting from zero and increasing according to an arithmetical proportion (or according to the series of natural numbers), what is the ratio of the sum of such a series to the sum of the same number of terms each equal to the last " ? †. In other words, what is the value of the ratio

$$\frac{0+1+2+3+4+\ldots+n}{n+n+n+n+n+\ldots+n} ?$$

This is the starting-point of an investigation which not only led to a method by which the quadrature of the circle could be effected but which also contained the germ of nearly every mathematical discovery of the next two centuries. The Method of Induction, which in his hands proved such a powerful instrument, is introduced on the very threshold of the treatise. "The simplest method of investigation", he declared in Proposition 1, "in this and in a certain number of problems which follow is to consider a certain number of individual cases, and to observe the emergent ratios, and to compare these with one another, so that a universal proposition may by induction be established " ‡. It is perhaps hardly necessary to mention that Wallis used the term Induction in the Baconian sense—namely, a generalization from a number of cases which would prove true universally. His use of the term bears but little resemblance to the so-called Mathematical Induction of the modern textbook. Thus, in order to obtain the ratio sought in Proposition 1, he notes that

$$\frac{0+1}{1+1} = \frac{1}{2},$$

$$\frac{0+1+2}{2+2+2} = \frac{3}{6} = \frac{1}{2},$$

* Dedication. "If therefore we can ascertain what is the ratio of the diameters of the former to those of the latter, we shall have found what we are seeking, since the former make up the Circle, and the latter make up the square."

† "Si proponatur series Quantitatum *Arithmetice-Proportionalium* (sive juxta naturalem numerorum consecutionem) continue crescentium, a puncto vel 0 (ciphra, seu nihilo) inchoatarum, (puta ut 0, 1, 2, 3, 4, etc.) propositum sit inquirere, quam habeat rationem earum omnium aggregatum, ad aggregatum totidem maximæ æqualium." Prop. 1.

‡ " Simplicissimus investigandi modus, in hoc et sequentibus aliquot Problematis, est, rem ipsam aliquousque præstare, et rationes prodeuntes observare atque invicem comparare; ut inductione tandem universalis propositio innotescat."

CHAPTER IV

$$\frac{0+1+2+3}{3+3+3+3} = \frac{6}{12} = \frac{1}{2},$$

$$\frac{0+1+2+3+4}{4+4+4+4+4} = \frac{10}{20} = \frac{1}{2}.$$

Proceeding in this way, he concludes that this ratio will persist, no matter what the number of terms may be, *i. e.*,

$$\frac{0+1+2+3+4+\ldots+n}{n+n+n+n+n+\ldots+n}$$

will always be equal to $\frac{1}{2}$.

Having established the truth of this proposition, Wallis next showed that, as a consequence, the area of a triangle or the volume of a paraboloid could be very easily determined. For consider the case of the triangle. The triangle, he declares (Proposition 3, Cor.), is made up of an infinite number of parallel lines arithmetically proportional, starting from zero, therefore the sum of the lines making up the triangle bears to the sum of those making up the rectangle on the same base and of the same altitude a ratio equal to $\frac{1}{2}$, since this ratio can manifestly be written

$$\frac{0+1+2+3+4+\ldots+n}{n+n+n+n+n+\ldots+n}.$$

Moreover, this ratio persists if n be infinite, *i. e.*, if the aggregate of the lines becomes identical with the area of the triangle ABC. Whence emerges the result that the area of the triangle is one-half that of the rectangle (fig. 3).

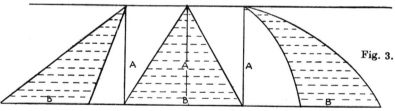

Fig. 3.

Similarly (Proposition 4, Cor.), the volume of the paraboloid is shown to be one-half the volume of the circumscribing cylinder; for since DO^2 is proportional to the area of the circle with DO as radius, and since DO^2 has already been proved (*Conic Sections*, Prop. 9), proportional to AD, then the volumes of the infinitely small discs which make up the paraboloid are proportional to AD, so that the ratio sought is reduced to the known ratio :

$$\frac{0+1+2+3+4+\ldots+n}{n+n+n+n+n+\ldots+n} = \frac{1}{2}. \qquad \text{[Fig. 4.]}$$

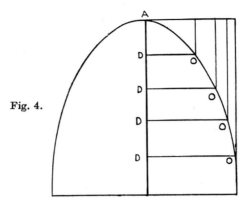

Fig. 4.

In Proposition 19 the investigation is directed to the ratio of the sum of the squares of numbers arithmetically proportional to the sum of an equal number of terms each equal to the greatest, and following the Method of Induction already established it is noted that

$$\frac{0+1}{1+1} = \frac{1}{2} = \frac{1}{3} + \frac{1}{6},$$

$$\frac{0+1+4}{4+4+4} = \frac{5}{12} = \frac{1}{3} + \frac{1}{12},$$

$$\frac{0+1+4+9}{9+9+9+9} = \frac{14}{36} = \frac{1}{3} + \frac{1}{18},$$

$$\frac{0+1+4+9+16}{16+16+16+16+16} = \frac{30}{80} = \frac{1}{3} + \frac{1}{24},$$

$$\frac{0+1+4+9+16+25}{25+25+25+25+25+25} = \frac{55}{150} = \frac{1}{3} + \frac{1}{30},$$

and generally

$$\frac{0^2+1^2+2^2+3^2+\ldots+n^2}{n^2+n^2+n^2+n^2+\ldots+n^2} = \frac{1}{3} + \frac{1}{6n},$$

whence Wallis concluded that the required ratio was rather more than 1/3, and that the excess over 1/3 diminished as the number of terms constituting the series increased. From this it was an easy step to show that if the number of terms were infinite the ratio sought would be 1/3. Pursuing this train of thought, Wallis was able to show the volume of a cone (or of a pyramid) to be 1/3 the circumscribing cylinder (or prism), and the area of a part

CHAPTER IV

of a parabola to be two-thirds its circumscribing rectangle. For in the latter case the ratio of the area of the small rectangles (which make up the area of the complement of the parabola) to that of the whole rectangle is easily shown to be

$$\frac{0^2+1^2+2^2+3^2+.+.+n^2}{n^2+n^2+n^2+n^2+.+.+n^2},$$

which ratio has already been shown to be equal to 1/3, whence the ratio of the area of the semi-parabola to that of the circumscribing rectangle is clearly 2/3 (fig. 5).

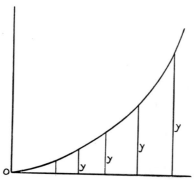

Fig. 5.

The case of the series of cubes of numbers arithmetically proportional is similarly dealt with at Proposition xxxix, where it is shown that

$$\frac{0+1}{1+1} = \frac{1}{2} = \frac{1}{4} + \frac{1}{4},$$

$$\frac{0+1+8}{8+8+8} = \frac{9}{24} = \frac{1}{4} + \frac{1}{8},$$

$$\frac{0+1+8+27}{27+27+27+27} = \frac{36}{108} = \frac{4}{12} = \frac{1}{4} + \frac{1}{12},$$

$$\frac{0+1+8+27+64}{64+64+64+64+64} = \frac{100}{320} = \frac{5}{16} = \frac{1}{4} + \frac{1}{16},$$

and generally

$$\frac{0^3+1^3+2^3+3^3+\ldots+n^3}{n^3+n^3+n^3+n^3+\ldots+n^3} = \frac{1}{4} + \frac{1}{4n}.$$

Whence by "Induction", as in Proposition 19, it is plain that this ratio for an infinite number of terms becomes 1/4.

These results lead to an obvious generalization, namely, that if for the cubes of numbers arithmetically proportional we substitute their 4th, 5th, 6th, etc. powers, the required ratio (which henceforth we shall call the *corresponding ratio*) becomes 1/5, 1/6, 1/7, etc. These results are summarized in Proposition 44, which he enunciates thus :

"If an infinite series of quantities be understood, beginning at 0 and continually increasing in arithmetical proportion (which series I call a series of *Laterales* or *Primana*) or of their Squares, Cubes, Fourth powers etc. (which series I shall call *Secundana*, *Tertiana*, *Quartana*, etc.), then the ratio of the sum of the whole series to the sum of a series of so many terms each equal to the greatest will be as is indicated in the following Table * :—

$n=0$.	Æquales.	1/1.		1.
$n=1$.	Primana.	1/2.		2.
$n=2$.	Secundana.	1/3.		3.
$n=3$.	Tertiana.	1/4.		4.
$n=4$.	Quartana.	1/5.	or as 1 to	5.
$n=5$.	Quintana.	1/6.		6.
$n=6$.	Sextana.	1/7.		7.
$n=7$.	Septimana.	1/8.		8.
$n=8$.	Octavana.	1/9.		9.

and so on."

The meaning of this is clear. As the indices of the terms constituting the several series form a progression whose terms increase by unity, so also do the consequents of the corresponding ratios form a like progression, the antecedent in each case being unity.

Before proceeding further with the investigation Wallis shows how the formulæ he has evolved will enable him to evaluate the volumes of such solid figures as the paraboloid etc. (Prop. 45).

It was a characteristic of Wallis, and one which distinguished him from every other mathematician of even that restless age, save Newton, that he always strove after generalized results, and it was through his attempts to generalize the results which he had already obtained that he was led to an interpretation of the meaning of negative and of fractional indices. He did not actually use either fractional or negative indices; still less did he write a function

* "Si intelligatur series infinita quantitatum, a puncto seu 0 inchoatarum, et continue crescentium pro ratione vel Arithmetice-proportionalium, (quam seriem *Lateralium* sive *Primanorum* appello) vel eorum quadratorum, cuborum, biquadratorum etc. (quam appello seriem *Secundanorum*, *Tertianorum*, *Quartanorum*, etc.). Erit totius seriei ratio, ad seriem totidem maximæ æqualium, ea quæ sequitur in hac Tabella, nempe."

such as $\sqrt[q]{a^p}$ as $a^{p/q}$. Nor was he the first to suggest the possible use of exponents other than positive integers. The idea of such had already been hinted at during the preceding century by Oresmes, Chuquet and Stevin, and, after all, what are logarithms but fractional and negative indices ? But in writing "$\dfrac{1}{x}$ cujus index est -1", "$\sqrt{x}$ cujus index est $\tfrac{1}{2}$", Wallis not only gave a definite meaning to such indices, but also prepared the way for Newton, to whom the modern notation is due.

The steps by which he extended his investigation to the case of fractional indices are introduced at Proposition 47, and his argument, briefly, is as follows :—

It has already been shown that as the number of terms increases, the ratio of the sum of the series

$$0^n + 1^n + 2^n + 3^n +$$

to the sum of an equally numerous series of terms, each equal to the last, continually approaches $\dfrac{1}{n+1}$. That is, as n assumes the values 1, 2, 3, 4, 5, etc. the consequents of the corresponding ratios of the different series form the similar progression

$$2, 3, 4, 5, 6, \text{etc.}$$

Now, asks Wallis, what must be the value of n if this ratio is to be 1 : 1 ? Clearly, it can only be zero. Hence the series whose corresponding ratio is 1 : 1 must be

$$0^0 + 1^0 + 2^0 + 3^0 + 4^0 +,$$

and this series he calls a series of *Æquales*. Further, says Wallis, the rule which has been shown to be true above will be no less valid if instead of starting from a series of *Primana* (*i. e.*, a series whose index is 1) we start with any other series, *e. g.*, one whose index is 2, 3, or 4, etc., and form from any one of these, new series whose terms are respectively the *squares* or the *cubes* etc. of the terms of the initial series*. Thus if we start with a series of squares (*Secundana*) whose corresponding ratio is 1 : 3, and form from it other series consisting of terms which are respectively the *squares*, or the *cubes* of these squares, we shall obtain other series whose

---

* This important step is enunciated in Prop. 47. " Hæc regula non minus valebit si exponatur series quantitatum quarumlibet (non quidem juxta seriem Primanorum, sed) juxta quamvis aliam Tabellæ seriem, et de illarum Quadratis, Cubis, etc. inquiratur."

corresponding ratios are 1 : 5 or 1 : 7, etc. That is to say, if we write the series

$0^0+1^0+2^0+3^0+$ (C.R.1 : 1),
$0^2+1^2+2^2+3^2+$ (C.R.1 : 3),
$0^4+1^4+2^4+3^4+$ (C.R.1 : 5),
$0^6+1^6+2^6+3^6+$ (C.R.1 : 7),

we notice at once that the arithmetical proportionality of the consequents of the corresponding ratios still persists. Now the converse of this is equally true. Thus, if we start with a series of *quartana*, $\quad 0^4+1^4+2^4+3^4+$

(whose corresponding ratio is 1 : 5), the consequent of the ratio of the series $\quad 0^2+1^2+2^2+3^2+,$

whose terms are respectively the *square roots* of the terms of the preceding series, is 3, and this is the arithmetic mean between 1 and 5, *i. e.*, between the consequents of the corresponding ratios of the series $\quad 0^0+1^0+2^0+3^0+$
$0^4+1^4+2^4+3^4+.$

To make Wallis's point clear, let us consider the series

$$0+1+8+27+64+$$

(whose corresponding ratio has been shown to be 1 : 4); then the series $\quad 0+1+2+3+4$

may be regarded as consisting of terms which are respectively the *cube roots* of the initial series. Similarly, the series

$$0+1+4+9+16+$$

is made up of terms which are respectively the *squares* of these cube roots. But it is noticed that these two derived series have corresponding ratios whose consequents are 2 and 3, numbers which may be regarded as having been obtained by the interpolation of two arithmetic means between 1 and 4. Similarly, had we started with the series $\quad 0+1+16+81+256+$

(corresponding ratio 1 : 5), then by taking the *fourth roots* of these, and the *squares of these fourth roots*, and the *cubes of these fourth roots*, we get three other series, thus :

Fourth roots .... $0+1+2+3+4+,$
Squares of fourth roots .... $0+1+4+9+16+,$
Cubes of fourth roots .... $0+1+8+27+64+.$

Now the corresponding ratios of these ratios have as their consequents the numbers 2, 3, 4. But what are 2, 3, 4 other than the three arithmetic means between 1 and 5 ?

CHAPTER IV

Having established this continuity, or principle of "interpolation" as Wallis calls it, the way to fractional indices is clear. For if we now consider the series

$$\sqrt{0}+\sqrt{1}+\sqrt{2}+\sqrt{3}+\sqrt{4}+,$$

whose terms are respectively the square roots of the terms of the series

$$0+1+2+3+4+,$$

then, by Proposition 53, the consequent of the corresponding ratio of this series ought to be the arithmetic mean between the consequents of the ratios of the other two series, namely, of

$$0^0+1^0+2^0+3^0+4^0+$$
and $$0^1+1^1+2^1+3^1+4^1+,$$

series whose ratios have consequents 1 and 2. Thus the series of square roots ought to have a corresponding ratio $1:1\frac{1}{2}$ (*conveniet ratio 1 ad $1\frac{1}{2}$, (sive 2 ad 3;) quia 1, $1\frac{1}{2}$, 2 sunt arithmetice proportionalia*, Prop. 53). Moreover, since the index of the terms of the series and the consequents of the corresponding ratios are related terms of the two arithmetical progressions, namely,

(indices) 0, 1, 2, 3, 4,
(ratios)   1, 2, 3, 4, 5, etc.,

it becomes plain that if we interpolate a term between 1 and 2 (*i. e.*, $1\frac{1}{2}$) in the lower row, we must interpolate a corresponding index between 0 and 1 (*i. e.*, $\frac{1}{2}$) in the upper row. In other words, the series

$$\sqrt{0}+\sqrt{1}+\sqrt{2}+\sqrt{3}+\sqrt{4}+$$

might just as well have been written

$$0^{\frac{1}{2}}+1^{\frac{1}{2}}+2^{\frac{1}{2}}+3^{\frac{1}{2}}+4^{\frac{1}{2}}+,$$

i. e., $\sqrt{x}$ is $x^{\frac{1}{2}}$, " cujus index est $\frac{1}{2}$ ".

Similarly, if we consider the series

$$0+1+2+3+4+$$

(Corresponding ratio 1 : 2), and from it derive the series

$$\sqrt[3]{0}+\sqrt[3]{1}+\sqrt[3]{2}+\sqrt[3]{3}+\sqrt[3]{4}+$$
and $$\sqrt[3]{0^2}+\sqrt[3]{1^2}+\sqrt[3]{2^2}+\sqrt[3]{3^2}+\sqrt[3]{4^2}+$$

then the consequents of these derived series must be the two arithmetic means between 1 and 2, *i. e.*, $1\frac{1}{3}$ and $1\frac{2}{3}$ ("quia scilicet 1, $1\frac{1}{3}$, $1\frac{2}{3}$, 2, sunt Arithmetice-proportionalia", Prop. 53), and thus

by the argument already advanced these latter series might have been written
$$0^{1/3}+1^{1/3}+2^{1/3}+3^{1/3}+$$
and
$$0^{2/3}+1^{2/3}+2^{2/3}+3^{2/3}+,$$
i. e., $\sqrt[3]{x}$ is $x^{1/3}$, and $\sqrt[3]{x^2}$ is $x^{2/3}$.

Proceeding in this way Wallis displays his results in a table (Proposition 54), which he explained thus:—

" If an infinite series of quantities be understood, starting from 0, and continually increasing according to the ratio of the square roots, cube roots, fourth roots, etc. of numbers arithmetically proportional (which I shall call series of *Subsecundana*, *Subtertiana*, *Subquartana* etc.), the ratio of the sum of the whole to the sum of an equal number of terms each equal to the greatest will be as is indicated in the following Table " *:

$n=1/2$. Subsecundana. $2/3$⎫ ⎧$1\frac{1}{2}$.
$n=1/3$. Subtertiana. $3/4$ ⎪ ⎪$1\frac{1}{3}$.
$n=1/4$. Subquartana. $4/5$ ⎬ or 1 to ⎨$1\frac{1}{4}$.
$n=1/5$. Subquintana. $5/6$ ⎪ ⎪$1\frac{1}{5}$.
$n=1/6$. Subsextana. $6/7$⎭ ⎩$1\frac{1}{6}$.

In other words, the above list gives the corresponding ratio of the series
$$0^n+1^n+2^n+3^n+4^n+,$$
when n successively assumes the values 1/2, 1/3, 1/4, 1/5, and so on.

The generalization of these results is at once clear.

(1) A number of the form $\sqrt[a]{r}$ can be expressed as a power of the form $r^{1/a}$.

(2) The corresponding ratio of the series
$$0^n+1^n+2^n+3^n+4^n+$$
is $\dfrac{1}{n+1}$, and this is shown to persist even if n is a fraction. In fact this latter result is not less valid if n be a fraction of the type p/q, i. e., if the series be
$$0^{p/q}+1^{p/q}+2^{p/q}+3^{p/q}+,$$
and Wallis's next step is to find the value which this corresponding ratio assumes for different values of p and q. Having evaluated these results, he then exhibits them in a Table (Prop. lix).

* " Si intelligatur series infinita quantitatum, a puncto seu 0 inchoatarum, et continue crescentium pro ratione Radicum quadraticarum, cubicarum, biquadraticarum etc., numerorum Arithmetice proportionalium; (quam appello seriem *Subsecundanorum*, *Subtertianorum*, *Subquartanorum etc.*). Erit totius ratio ad seriem totidem maximæ æqualium, ea quæ sequitur in hac Tabella: Nempe. (Prop. liv.)

CHAPTER IV

TABLE I.
Values of the Corresponding Ratios of the Series $0^{p/q} + 1^{p/q} + 2^{p/q} + 3^{p/q} + 4^{p/q} + \ldots$ for different Values of p and q.

	Series (*i.e.*, different values of p).									
Roots (*i.e.*, different values of q).	Æqualium. $p=0.$	Primanorum. $p=1.$	Secundanorum. $p=2.$	Tertianorum. $p=3.$	Quartanorum. $p=4.$	Quintanorum. $p=5.$	Sextanorum. $p=6.$	Septimanorum. $p=7.$	Octavanorum. $p=8.$	Nonanorum. $p=9.$
Quadraticæ, $q=2$	$\frac{1}{1}$	$\frac{1}{2}$	$\frac{1}{3}$	$\frac{1}{4}$	$\frac{1}{5}$	$\frac{1}{6}$	$\frac{1}{7}$	$\frac{1}{8}$	$\frac{1}{9}$	$\frac{1}{10}$
Cubicæ, $q=3$	$\frac{2}{2}$	$\frac{2}{3}$	$\frac{2}{4}$	$\frac{2}{5}$	$\frac{2}{6}$	$\frac{2}{7}$	$\frac{2}{8}$	$\frac{2}{9}$	$\frac{2}{10}$	$\frac{2}{11}$
Biquadrat., $q=4$	$\frac{3}{3}$	$\frac{3}{4}$	$\frac{3}{5}$	$\frac{3}{6}$	$\frac{3}{7}$	$\frac{3}{8}$	$\frac{3}{9}$	$\frac{3}{10}$	$\frac{3}{11}$	$\frac{3}{12}$
Sursolidæ, $q=5$	$\frac{4}{4}$	$\frac{4}{5}$	$\frac{4}{6}$	$\frac{4}{7}$	$\frac{4}{8}$	$\frac{4}{9}$	$\frac{4}{10}$	$\frac{4}{11}$	$\frac{4}{12}$	$\frac{4}{13}$
Sextanæ, $q=6$	$\frac{5}{5}$	$\frac{5}{6}$	$\frac{5}{7}$	$\frac{5}{8}$	$\frac{5}{9}$	$\frac{5}{10}$	$\frac{5}{11}$	$\frac{5}{12}$	$\frac{5}{13}$	$\frac{5}{14}$
Septiminæ, $q=7$	$\frac{6}{6}$	$\frac{6}{7}$	$\frac{6}{8}$	$\frac{6}{9}$	$\frac{6}{10}$	$\frac{6}{11}$	$\frac{6}{12}$	$\frac{6}{13}$	$\frac{6}{14}$	$\frac{6}{15}$
	$\frac{7}{7}$	$\frac{7}{8}$	$\frac{7}{9}$	$\frac{7}{10}$	$\frac{7}{11}$	$\frac{7}{12}$	$\frac{7}{13}$	$\frac{7}{14}$	$\frac{7}{15}$	$\frac{7}{16}$

But Wallis's "Induction" did not stop here. Having shown the ratio $\frac{1}{n+1}$ to be true for positive values of n, including fractions, with rare boldness he now assumes it to be true even if n is irrational! "Sin Index supponatur irrationalis, puta $\sqrt{3}$, erit ratio ut 1 ad $1+\sqrt{3}$", *i. e.*, if $n=\sqrt{3}$, the emergent corresponding ratio is $\frac{1}{1+\sqrt{3}}$. (Prop. lxiv.)

This exhausts the investigation of the cases in which the indices, though no longer integral, are nevertheless positive. With a prolixity which was perhaps understandable (for he was on virgin soil), Wallis embarks at Proposition 73 into the region of negative indices. His steps are:

(1) If two series are multiplied together, term by term, we shall get a third series, the index of whose terms is the sum of the indices of the two initial series; moreover, the consequent of the corresponding ratio of this derived series will be such as is indicated in the appropriate space in the preceding table, *i. e.*, it will be the sum of the consequents of the ratios of the two individual series, diminished by one. Further, this will persist whether the indices under consideration are integral or fractional. Thus, to take Wallis's own example, if a series of *secundana* (whose index is 2) be multiplied, term by term, by a series of *subtertiana* (whose index is 1/3), there will emerge a series of *radices-cubicæ-septimanorum*, whose index is 7/3; moreover, the corresponding ratio of this emergent series will be $\frac{1}{1+7/3}$, or $\frac{3}{10}$.

Or again, if the series

$$0^{\frac{1}{2}}+1^{\frac{1}{2}}+2^{\frac{1}{2}}+3^{\frac{1}{2}}+4^{\frac{1}{2}}+$$

be multiplied, term by term, by the series

$$0^{1/5}+1^{1/5}+2^{1/5}+3^{1/5}+4^{1/5}+$$

we shall get the series

$$0^{7/10}+1^{7/10}+2^{7/10}+3^{7/10}+4^{7/10}+,$$

a series whose corresponding ratio is $\frac{1}{1+7/10}$ or $\frac{10}{17}$.

(2) If a series be divided by another series, term by term, a third series will emerge, the index of whose terms will be the difference of the indices of the two original series. "If all the terms of one series", says Wallis, "be respectively divided by the terms of a second series, the quotients will form another series whose

CHAPTER IV 41

index is found by subtracting the index of the divisor from that of the dividend; the difference is the index of the series of quotients " *. Thus a series of *quartana*, e. g.,

$$0^4 + 1^4 + 2^4 + 3^4 +$$

divided by a series of *tertiana*

$$0^3 + 1^3 + 2^3 + 3^3 +,$$

gives a series of *Primana* (*Laterales seu Primana, cujus index* $1 = 4 - 3$). If, however, the index of the terms of the second series be greater than that of the former, this rule will yield a negative result, *e. g.*, if a series of squares be divided by a series of cubes we get, by the above rule, a series of terms whose index is $2-3$, or -1. But if we divide, term by term, a series of squares, *e. g.*,

$$0a^2 + 1a^2 + 4a^2 + 9a^2 + 16a^2$$

by a series of cubes, *e. g.*,

$$0a^3 + 1a^3 + 8a^3 + 27a^3 + 64a^3,$$

we get the series

$$\frac{1}{0a} + \frac{1}{1a} + \frac{1}{2a} + \frac{1}{3a} + \frac{1}{4a},$$

whence it becomes plain that $\frac{1}{a}$ is the same as a^{-1}, and similarly, $\frac{1}{a^2}$ is the same as a^{-2}, i. e., $\frac{1}{a^n}$ is the same as a^{-n}. "Let the series thus emerging be called Reciprocal Series, and let them have negative Indices " †.

All this, of course, was a great advance upon anything that had gone before, and Wallis well knew that it was. Even at the risk of prolixity he preferred, as he said, to throw open the very fount to the reader, rather than imitate the methods of the ancients, who strove to be admired than be understood. In Proposition 106, therefore, he summarized his investigation upon fractional and negative indices thus :

"If any reciprocal series be multiplied by another (either reciprocal or direct), or divided, or even if it multiply or divide the same, the laws are to be observed as in the direct series (Props.73

* "Si unius Seriei termini omnes per terminos alterius Seriei respective dividantur, quotientes erunt series alia, cujus index reperitur subducendo indicem seriei Dividentis ex indice seriei divisæ, quod restat enim est Index seriei divisione provenientis, sive Quotientis." (Prop. lxxxi.)

† "Series autem sic provenientes, series *Reciprocæ* appellentur, habeantque Indices negativos." (Prop. lxxxvii.)

and 81); *e.g.*, if the series of reciprocals of the squares (*e. g.*, $\frac{1}{1}, \frac{1}{4}, \frac{1}{9}$), whose index is -2 be multiplied term by term into the series of reciprocals of cubes (whose index is -3), there will emerge a series of *subquintana* (reciprocals of the fifth powers) $\frac{1}{1}, \frac{1}{32}, \frac{1}{243}$, etc., whose index $-5=-2-3$, as is evident.

" Furthermore, if a series of reciprocals of cubes ($\frac{1}{1}, \frac{1}{8}, \frac{1}{27}$), whose index is -3, is multiplied term by term by a series of squares (*secundana*) 1, 4, 9, etc., whose index is 2, there will emerge the series $\frac{1}{1}, \frac{4}{8}, \frac{9}{27}, \ldots$ *i. e.*, $\frac{1}{1}, \frac{1}{2}, \frac{1}{3}$; .... a series of reciprocals of first powers whose index is $-1=-3+2$.

" Likewise if the series of reciprocals of the square roots (*subsecundana*), $\frac{1}{\sqrt{1}}, \frac{1}{\sqrt{2}}, \frac{1}{\sqrt{3}}, \ldots$ whose index is $-\frac{1}{2}$, be multiplied term by term, by a series of squares (*secundana*) 1, 4, 9, .. whose index is 2, the products will form the series

$$\frac{1}{\sqrt{1}}, \frac{4}{\sqrt{2}}, \frac{9}{\sqrt{3}} \ldots,$$

or
$$\frac{1\sqrt{1}}{1}, \frac{4\sqrt{2}}{2}, \frac{9\sqrt{3}}{3},$$

or
$$\sqrt{1}, \sqrt{8}, \sqrt{27},$$

the square roots of the cubes or third powers, whose index is $\frac{3}{2}=-\frac{1}{2}+2$.

" Furthermore, if a series of reciprocals of squares whose index is -2 be divided into a series of reciprocals of integers (*Primana*) whose index is -1, the resultant series will be a series of first powers (*Primana*) whose index $1=-1+2$, or -1 minus -2.

" Likewise, if a series of reciprocals of integers, (*Primana*) whose index is -1, is divided into a series of reciprocals of squares (*secundana*) whose index is -2, the resultant series will be a series of reciprocals of first powers whose index is $-1=-2$ minus -1.

" Likewise, if a series of reciprocals of first powers whose index is -1 is divided into a series of squares whose index is 2, the resultant series will be a series of third powers (*Tertiana*), whose index is $3=2+1$, or 2 minus -1.

CHAPTER IV

"Likewise if a series of reciprocals of first powers whose index is -1 is divided by a series of squares whose index is 2, the resultant series will be a series of reciprocals of cubes, whose index $-3=-1-2$, or -1 minus 2.

"And the same thing will occur in any other series whatsoever of this kind, and thus the proposition is proved."

But here a formidable obstacle presented itself, which not even the genius of Wallis was able to overcome. Wallis had shown that the corresponding ratio of the series

$$0^n + 1^n + 2^n + 3^n + 4^n +$$

was $\dfrac{1}{n+1}$, and that this ratio persisted even if n were fractional. Moreover, this had led to the determination of the areas defined by curves in which one co-ordinate was proportional to a power of the other, that power of course being positive. But when he tried to extend his investigations to the case where the curve was such that one co-ordinate was proportional to the reciprocal of the other, or a higher power of the reciprocal, the result which emerged was to him unintelligible. For, according to the law of continuity which he had established, the ratio of the area under a curve of this type to the area of its circumscribing rectangle should be expressed by the formula $\dfrac{1}{n+1}$ even if n were negative. In the hyperbola $y = \dfrac{1}{\sqrt{x}}$, in which $n = -\tfrac{1}{2}$, the area under the curve bears to the circumscribing rectangle the ratio $1 : -\tfrac{1}{2}+1$, or $1 : \tfrac{1}{2}$. If $n=-1$, as in the hyperbola $y=1/x$, this ratio becomes $1/0$, which, it will be recalled, he had already in the *Conic Sections* and elsewhere characterized as *Infinite*. If n were numerically greater than 1, i. e., if the curve were of the type $y=1/x^2$, $1/x^3$, etc., the consequent of the emergent ratio would be negative. But to him the ratio of a positive quantity to a negative one was meaningless, and his attempts to elucidate it led to the curious conclusion that these ratios were *greater than infinite*! This result was reached by a study of the two sets of series :

$$0^1,\ 1^1,\ 2^1,\ 3^1,\ 4^1,$$
$$0^2,\ 1^2,\ 2^2,\ 3^2,\ 4^2,$$
$$0^3,\ 1^3,\ 2^3,\ 3^3,\ 4^3,$$

and

$$\frac{1}{0},\ \frac{1}{1},\ \frac{1}{2},\ \frac{1}{3},\ \frac{1}{4},$$

$$\frac{1}{0^2}, \frac{1}{1^2}, \frac{1}{2^2}, \frac{1}{3^2}, \frac{1}{4^2},$$

$$\frac{1}{0^3}, \frac{1}{1^3}, \frac{1}{2^3}, \frac{1}{3^3}, \frac{1}{4^3},$$

and so on *.

Now let us consider the value of the corresponding ratio of each of the above series, i. e., the corresponding ratio of the series formed by giving n successive values, say from -4 to $+4$ in the series

$$0^n + 1^n + 2^n + 3^n + .$$

The value of the corresponding ratio will be

$n = 4.$ Ratio $= 1/5.$
$n = 3.$,, $= 1/4.$
$n = 2.$,, $= 1/3.$
$n = 1.$,, $= 1/2.$
$n = 0.$,, $= 1/1.$
$n = -1.$,, $= 1/0.$
$n = -2.$,, $= 1/-1.$
$n = -3.$,, $= 1/-2.$
$n = -4.$,, $= 1/-3.$

Clearly these ratios increase from the first to the sixth, i. e., from $1/5$ to $1/0$; might we not infer, therefore, that they continue to increase even beyond $1/0$, in which case we may write

$$\frac{1}{0} > \frac{1}{-1} > \frac{1}{-2} > \frac{1}{-3} > \frac{1}{-4}.$$

But $\frac{1}{0}$ has already been shown to be infinite, therefore, argues

* Nowhere in the whole of his writings is the conception of negative exponents more clearly indicated than it is in this passage: "Ubi autem series directæ indices habent 1, 2, 3, etc., ut quæ supra seriem Æqualium tot gradibus ascendunt; habebunt hæ quidem (illis reciprocæ) suos indices contrarios negativos $-1, -2, -3$, etc., tanquam tot gradibus infra seriem Æqualium descendentes. Prout autem illæ ab 0 ciphra vel Nihilo continuo crescunt, hæ contra ab ∞ Infinito continue decrescunt; atque ut illic terminus maximus, hic minimus, seriem claudit", i. e., where the direct series have indices, 1, 2, 3, etc., continually increasing by steps above the series of *Æquales*, the second (their reciprocals), on the other hand, will have negative indices successively decreasing as $-1, -2, -3$, etc.; as the former increase from zero, so do the latter decrease from ∞ Infinity; in the former the greatest term closes the series, in the latter it is the least. (Prop. ci, Scholium.)

Wallis, the ratios $\dfrac{1}{-1}, \dfrac{1}{-2}$, etc. must be *greater than infinite*. Thus Wallis arrived at a position which he did not understand, namely, the transition from a positive to a negative quantity by way of infinity. His error lay in the fact that he assumed $\dfrac{1}{a}$ to increase continually as a by units diminished, and that this increase persisted when $a=0$. Further, since the formula for the corresponding ratio gives a measure of the area under the curve of the type $y=x^n$, it follows, asserted Wallis, that a curve defined by co-ordinates such that the ordinate is proportional to the reciprocal of the square, or cube, etc. of the abscissa will have an area greater than infinity ! " A ratio greater than Infinity, such as a positive number

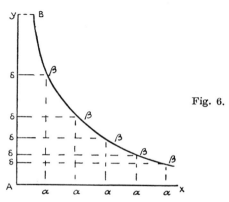

Fig. 6.

may be supposed to have to a negative number, or one less than nothing " *.

These conclusions are illustrated geometrically. In the curve shown (fig. 6) the ordinates such as $\alpha\beta$ are inversely proportional to the abscissæ $\delta\beta$. Any ordinate $\alpha\beta$ to the curve being drawn, what is the ratio of the area bounded by $\alpha\beta$, the axes, and the curve, to that of the rectangle $A\beta$? If we erect ordinates at unit distances from A, we shall get, using the rule already proved, a measure of this ratio, for the ordinates so erected at distances 0, 1, 2, 3, 4, etc. will form the terms of the series

$$0^{-1},\ 1^{-1},\ 2^{-1},\ 3^{-1},\ 4^{-1},$$

* "Rationem plusquam infinitam ; qualem nempe habere supponatur numerus positivus ad numerum negativum, sive minorem nihilo." (Prop. civ.)

which, by the rule, gives the ratio sought as $\frac{1}{0}$ or infinity. "This will have to the inscribed parallelogram an infinite ratio namely, that which 1 is to 0 " *.

If, however, the curve is such that the ordinates decrease not as the reciprocals of the abscissæ, but as the squares or higher powers of these reciprocals, i. e., if the ordinates erected at successive unit intervals along AX are respectively

$$0^{-2},\ 1^{-2},\ 2^{-2},\ 3^{-2},\ 4^{-2},$$
or $$0^{-3},\ 1^{-3},\ 2^{-3},\ 3^{-3},\ 4^{-3},$$

then the ratio of the areas under consideration is $\frac{1}{-1}$ (or $\frac{1}{-2}$), and this is greater than infinity! " For when the indices of a series of squares, cubes etc.", says Wallis, " are 2, 3, 4, etc. (greater than unity), the indices of the series of reciprocals of the series will be -2, -3, etc., which although increased by unity, (according to Prop. 64) nevertheless remain negative, as $-2+1=-1$, $-3+1=-2$, $-4+1=-3$, etc., and thus the ratio which 1 bears to these indices thus increased, e. g., 1 to -1, 1 to -2, 1 to -3, etc. shall be greater than infinite, or 1 : 0 because the consequents of the ratios are less than 0 " †. It was left for Varignon to show that the numerical value of the ratio is merely a measure of the area of the space on the other side of the ordinate.

Wallis's next step is to show that his method of finding areas applies with equal validity to cases more complex, i. e., those in which the ordinate is equal to a compound expression, as, for example, in the curve $y=(a+x)^2$. Now it is evident that we can assume an ordinate of this type to be identical with the sum of several ordinates, of which one is constant, and thus, following the rules already established, the area under the curve of the type $y=(a+x)^2$, or $a^2+2ax+x^2$ will be the sum of several areas which are respectively equal to a^2x, ax^2, and $\frac{x^3}{3}$. From considerations such as these Wallis envisaged an ingenious way of effecting the quadrature of the circle, for if the ordinate of the circle could be

---

* " Habebit illa ad Parallelogrammum inscriptum rationem infinitam, eam nempe quæ est 1 ad 0." (Prop. ciii.)

† " Puta cum indices seriei secundanorum, Tertianorum, Quartanorum etc. sint 2, 3, 4, (unitate majores,) indices serierum illis reciprocarum erunt -2, -3, -4, etc. qui quamvis unitate augeantur (juxta Prop. 64) manebunt tamen negativi, puta $-2+1=-1$, $-3+1=-2$, $-4+1=-3$ etc., et propterea ratio quam habet 1 ad indices illos sic auctos, puta 1 ad -1, 1 ad -2, 1 ad -3 etc. major erit quam infinita, sive 1 ad 0, quia nempe rationum consequentes sunt minores quam 0." (Prop. civ.)

CHAPTER IV 47

expanded in terms of x its quadrature could be readily accomplished. Wallis, however, had no means of effecting such expansion, and therefore he had to fall back on other methods. Not until we reach Proposition 121 do we find any mention of the circle—a striking commentary upon the thoroughness with which he laid his foundations. In this proposition he reaches out to his main objective, namely, the determination of the ratio of the area of the quadrant of the circle to the square on the radius, *i. e.*, the determination of $\frac{\pi}{4}$.

If we consider the semicircle of radius R, whose diameter is divided into a large number of equal parts a, and equidistant perpendiculars are erected thereon, it is clear that since these perpendiculars are the mean proportionals of the segments of the diameter they will form the series

$$\sqrt{R^2-0a^2},\ \sqrt{R^2-1a^2},\ \sqrt{R^2-4a^2},\ \sqrt{R^2-9a^2},\text{ etc. [Fig. 7.]}$$

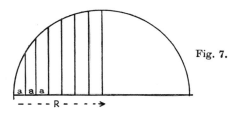

Fig. 7.

If, therefore, we can find the corresponding ratio of the series whose general term is $\sqrt{R^2-p^2a^2}$, we shall obtain the ratio of the area of the quadrant to that of the square on the radius, *i. e.*, $\frac{\pi}{4}$. Now it is obvious that a series whose general term may be written $(R^2-p^2a^2)^{\frac{1}{2}}$ can be regarded as occupying a position midway between the first and second members of the series whose general terms are

$$(R^2-p^2a^2)^0,\ (R^2-p^2a^2)^1,\ (R^2-p^2a^2)^2,\ (R^2-p^2a^2)^3,\text{ etc.},$$

and his first surmise was the not unnatural one that its corresponding ratio should lie midway between the corresponding ratios of these two members. This, however, leads to the value $\sqrt{\frac{2}{3}}$ for $\frac{\pi}{4}$, which makes π equal 3·26, a value which is clearly too large. Not only that, but such a method of interpolation was shown to be untenable when he proceeded to find the corresponding ratio of the other terms of the series. This was demonstrated at length by

Wallis. The corresponding ratio of the series whose general term is $(R^2-p^2a^2)^0$ is obviously $1:1$. To determine that for the other members, e. g. $(R^2-p^2a^2)^1$, Wallis goes back to the theorem he has already proved, namely, that if a series of *Æquales* be diminished by a series of *Secundana* the corresponding ratio of the emergent series will be $2/3$. (*Si series Æqualium mulctetur serie Secundanorum . . . . residua erunt totius duo trientes . . . .*, Prop. cxi.) To put it another way, if we sum the series

$$R^2-0a^2,$$
$$R^2-1a^2,$$
$$R^2-4a^2,$$
$$R^2-9a^2,$$
$$\cdot \ \cdot \ \cdot \ \cdot$$

to $\qquad R^2-p^2a^2,$

where $\qquad p^2a^2=R^2,$

we recognize two distinct series, namely,

$$R^2+R^2+R^2+R^2+$$

and $\qquad (0a)^2+(1a)^2+(2a)^2+(3a)^2+.$

The corresponding ratio of the former is clearly $1:1$, and that of the latter has already been shown to be $1:3$. Hence the corresponding ratio of the diminished series is $1-1/3$ or $2/3$.

In a similar way he shows (Prop. 118) that the corresponding ratio of the series whose general term is $(R^2-p^2a^2)^2$ to be $8/15$, for on summation of this series we get

$$R^4-2\times\ 0a^2+\ 0a^4,$$
$$R^4-2\times\ \ a^2+\ 1a^4,$$
$$R^4-2\times\ 4a^2+16a^4,$$
$$R^4-2\times\ 9a^2+81a^4,$$
$$\cdot \ \cdot \ \cdot \ \cdot \ \cdot \ \cdot \ \cdot$$

to $\qquad R^4-2\times p^2a^2+p^4a^4.$

Here we recognize three distinct series, namely,

$$R^4+R^4+R^4+R^4+\ldots\ (\text{C.R. } 1:1),$$
$$2\{0a^2+(1a)^2+(2a)^2+(3a)^2+\}\ldots\ (\text{C.R. } 1:3),$$
$$(0a)^4+(1a)^4+(2a)^4+(3a)^4+\ldots\ (\text{C.R. } 1:5),$$

and the corresponding ratio of this combined series is clearly $1-2/3+1/5$ or $8/15$, and this, he notes, may be written $\dfrac{2\times 4}{3\times 5}$.

CHAPTER IV

By a similar process of reasoning the corresponding ratio of the series whose general term is $(R^2-p^2a^2)^3$ is shown to be

$$1-3/3+3/5-1/7 \text{ or } \frac{48}{105} \text{ or } \frac{2\times 4\times 6}{3\times 5\times 7},$$

" and thus by continued multiplication of numbers arithmetically proportional (as many as is indicated by the index of the power) from 2 and 3, increasing by two's " *. Wallis was clearly striving after a general rule, and it looked almost as though he had succeeded in establishing one, namely, that the corresponding ratio of a series whose general term is $(R^2-p^2a^2)^n$ might be obtained by taking the continued product to n terms of the series

$$\frac{2\times 4\times 6\times}{3\times 5\times 7\times}.$$

But in the case sought, namely $(R^2-p^2a^2)^{\frac{1}{2}}$, n being equal to $\frac{1}{2}$ Wallis's difficulty is at once manifest. So now the original problem has assumed a new aspect, namely:

If in the series whose general terms are

$$(R^2-p^2a^2)^0, \ (R^2-p^2a^2)^1, \ (R^2-p^2a^2)^2, \ (R^2-p^2a^2)^3, \text{ etc.},$$

which have for their corresponding ratios

$$1, \quad 2/3, \quad 8/15, \quad 48/105,$$

a series whose general term is $(R^2-p^2a^2)^{\frac{1}{2}}$ is now interpolated between the first and second terms, what will be the corresponding ratio of this interpolated series? The answer to this question will give the quadrature of the circle †. But the answer to this question demanded a knowledge of the law of the series, and this was beyond Wallis's ken. His design now becomes a search, therefore, for a series in which interpolation of the term $(R^2-p^2a^2)^{\frac{1}{2}}$ is possible, and from which the value of this interpolated term might be determined.

His first step is to determine the corresponding ratios of the different series:

$(R-a)^0, \quad (R-a)^1, \quad (R-a)^2, \quad (R-a)^3, \quad$ etc.

$(\sqrt[2]{R}-\sqrt[2]{a})^0, \ (\sqrt[2]{R}-\sqrt[2]{a})^1, \ (\sqrt[2]{R}-\sqrt[2]{a})^2, \ (\sqrt[2]{R}-\sqrt[2]{a})^3, \quad$,,

$(\sqrt[3]{R}-\sqrt[3]{a})^0, \ (\sqrt[3]{R}-\sqrt[3]{a})^1, \ (\sqrt[3]{R}-\sqrt[3]{a})^2, \ (\sqrt[3]{R}-\sqrt[3]{a})^3, \quad$,,

$(\sqrt[4]{R}-\sqrt[4]{a})^0, \ (\sqrt[4]{R}-\sqrt[4]{a})^1, \ (\sqrt[4]{R}-\sqrt[4]{a})^2, \ (\sqrt[4]{R}-\sqrt[4]{a})^3, \quad$,,

$(\sqrt[5]{R}-\sqrt[5]{a})^0, \ (\sqrt[5]{R}-\sqrt[5]{a})^1, \ (\sqrt[5]{R}-\sqrt[5]{a})^2, \ (\sqrt[5]{R}-\sqrt[5]{a})^3, \quad$,,

---

* " Et sic deinceps, continue multiplicando numeros arithmetice proportionales, (quousque gradus potestatis postulat) a 2 et 3 continue binario crescentes." (Prop. cxviii.)

† "Si rationum series illa 1/1, 2/3, 8/15, 48/105 etc. interpolari poterit; ratio illa quæ primæ et secundæ ponenda est intermedia, est ea quam habet Circuli Quadrans ad Quadratum Radii, vel Circulus ipse ad quadratum Diametri." (Scholium, Prop. clxv.)

The determination of these corresponding ratios involves the expansion of the different series. Thus the members of the first row become

$$1,\ (R-a),\ (R^2-2Ra+a^2),\ (R^3-3R^2a+3Ra^2-a^3),$$

and their corresponding ratios are easily shown to be

$$1,\ 1-\tfrac{1}{2},\ 1-2/2+1/3,\ 1-3/2+3/3-1/4,\ \text{etc.},$$

or, $\qquad\qquad 1,\ 1/2,\ 1/3,\ 1/4.$

Similarly, the different series of the second row become on expansion:

$$1,\ R^{\frac{1}{2}}-a^{\frac{1}{2}},\ R-2R^{\frac{1}{2}}a^{\frac{1}{2}}+a,\ R^{3/2}-3Ra^{\frac{1}{2}}+3R^{\frac{1}{2}}a-a^{3/2},$$

and their corresponding ratios are shown to be

$$1,\ 1-2/3,\ 1-4/3+1/2,\ 1-6/3+3/2-2/5,\ \text{etc.},$$

i. e., $\qquad\qquad 1,\ 1/3,\ 1/6,\ 1/10,$

or $\qquad\qquad \dfrac{1}{1},\ \dfrac{1}{1+2},\ \dfrac{1}{1+2+3},\ \dfrac{1}{1+2+3+4},$

" and so on, that ratio which 1 bears to triangular numbers, or the aggregate of numbers arithmetically proportional, (as required by the degree of the power) continually increasing by unity " *.

The terms of the third row when expanded become

$$1,\ (R^{\frac{1}{3}}-a^{\frac{1}{3}}),\ (R^{\frac{2}{3}}-2R^{\frac{1}{3}}a^{\frac{1}{3}}+a^{\frac{2}{3}}),\ (R-3R^{\frac{2}{3}}a^{\frac{1}{3}}+3R^{\frac{1}{3}}a^{\frac{2}{3}}-a),$$

and their corresponding ratios are shown to be

$$1,\ 1-3/4,\ 1-6/4+3/5,\ 1-9/4+9/5-1/2,$$

i. e., $\qquad\qquad 1,\ 1/4,\ 1/10,\ 1/20,$

or $\qquad\qquad \dfrac{1}{0+1},\ \dfrac{1}{1+3},\ \dfrac{1}{4+6},\ \dfrac{1}{10+10}.$

In exactly the same way the corresponding ratios of the fourth and fifth rows are shown to be

$$1,\quad 1/5,\quad 1/15,\quad 1/35,\ \text{etc.}$$

and $\qquad 1,\quad 1/6,\quad 1/21,\quad 1/56,\ \text{,,}$

All these results Wallis exhibits in a table, in which are shown the consequents of the corresponding ratios of the series whose general term may be written $(\sqrt[p]{R}-\sqrt[p]{a})^n$.

* " Et sic deinceps; nempe eam rationem quam habet 1 ad numeros triangulares, sive aggregatum numerorum arithmetice proportionalium (quousque gradus potestatis postulat) ab 1 continue unitate crescentium." (Prop. cxxviii.)

CHAPTER IV

TABLE II. (Prop. 132.)

Consequents of the Corresponding Ratios of the Series whose General Term is $(\sqrt[p]{R}-\sqrt[p]{a})^n$.

		POWERS (i. e., values of n).							
		0.	1.	2.	3.	4.	5.	6.	7.
ROOTS (i. e., values of p).	0	1	1	1	1	1	1	1	1
	1	1	2	3	4	5	6	7	8	$(R-a)^n$.
	2	1	3	6	10	15	21	28	36	$(\sqrt[2]{R}-\sqrt[2]{a})^n$.
	3	1	4	10	20	35	56	84	120	$(\sqrt[3]{R}-\sqrt[3]{a})^n$.
	4	1	5	15	35	70	126	210	330	$(\sqrt[4]{R}-\sqrt[4]{a})^n$.
	5	1	6	21	56	126	252	462	792	$(\sqrt[5]{R}-\sqrt[5]{a})^n$.
	6	1	7	28	84	210	462	924	1716	$(\sqrt[6]{R}-\sqrt[6]{a})^n$.
	7	1	8	36	120	330	792	1716	3432	$(\sqrt[7]{R}-\sqrt[7]{a})^n$.

The table thus constructed exhibits certain distinctive features, to which Wallis calls attention in the Scholium to Proposition 132 and again in Proposition 169.

(1) The numbers in the rows are figurate. "Numeri omnes Tabellæ Prop. 132 sunt figurati. Nempe, qui in illius serie prima, (sive erecta sive transversa) sunt monadici qui in tertia, Triangulares, qui in quarta, Pyramidales; etc." (Prop. 169). That is, the numbers in the first row are monadici, those in the third are triangulars, those in the fourth row are pyramidal numbers, and so on *.

(2) The columns merely repeat the rows. (Note that he makes them identical by the insertion of the row 1, 1, 1, 1, corresponding to the terms of the series $(\sqrt[0]{R}-\sqrt[0]{a})^n$.

* *Note on Figurate Numbers.* Triangular numbers are formed by the addition of the terms of an arithmetic progression whose first term is 1, and whose common difference is 1. Their general formula is $n(n+1)/1.2$. The numbers formed by adding the triangular numbers are called *Pyramidals*, and their general formula is $\frac{n(n+1)(n+2)}{1.2.3}$. Similarly the numbers formed by adding these Pyramidals are called *Second Pyramidals* or *Triangulipyramidals*, and their general formula is $\frac{n(n+1)(n+2)(n+3)}{1.2.3.4}$.

These two facts led Wallis to note a principle whereby he could extend his table indefinitely. Moreover, he now directs attention to the fact that interpolation in this Table will lead to the evaluation of π, since, as he points out, "The Circle is to the square on the diameter as 1 is to □, the number interpolated between 1 and 2 in the diagonal of the foregoing Table" (Prop. clxviii, Cor.). This is obvious if we remember that the quantity sought, namely $(R^2-a^2)^{\frac{1}{2}}$, is identical with $(\sqrt[\frac{1}{2}]{R}-\sqrt[\frac{1}{2}]{a})^{\frac{1}{2}}$, and thus falls under the series whose general term is $(\sqrt[p]{R}-\sqrt[p]{a})^n$, in which p and n are each equal to $\frac{1}{2}$. Putting □ for the value of this (i. e., $\frac{4}{\pi}$), we can thus find a place for □ in the diagonal of the table. So the table is rewritten in the form shown in Table III, where the space occupied by □ is indicated. But if we make a gap for □ we are left with a number of other gaps to be filled, and Wallis proceeds to show how this can be accomplished.

TABLE III.

	Monadici.	Laterales.	Triangulares.	Pyramidales.	Triangular-Pyram.	Pyram-Pyramid.
Monadici	1	1	1	1	1	1
		□				
Laterales	1	2	3	4	5	6
Triangulares	1	3	6	10	15	21
Pyramidales	1	4	10	20	35	56
Triangulipyram	1	5	15	35	70	126
Pyramidipyram	1	6	21	56	126	252

CHAPTER IV

The rows are figurate, those in the first row being *Monadici*, those in the second row *Laterales*, in the third row *Triangulares*, those in the fourth row *Pyramidales*, and so on. Now in the first two rows, the Monadici and the Laterales, the law of succession is obvious; these rows can therefore be completed, and with them the corresponding columns.

To obtain the numbers in the third row (Triangulars), Wallis directs attention to the proposition: "Required the ratio of the triangular number to its side"*—that is, what is the triangular number whose side is n, and, of course, the answer is $\frac{n(n+1)}{2}$. The values corresponding to $n=1$, 2, 3 etc. agree with the numbers (1, 3, 6, 10, 15, etc.) in the odd columns; Wallis now perceives, using his favourite method of "induction", a means of filling in the even columns. Assuming the rule for triangular numbers to hold when $n=1$, 2, 3 etc., why should it not, he asks, persist for a number interpolated between these, *i.e.*, for 3/2? Putting $n=3/2$, he finds the corresponding triangular number to be $\frac{15}{8}$. Putting n successively equal to $\frac{1}{2}$, $\frac{5}{2}$, $\frac{7}{2}$, etc., he finds the value of the triangular numbers corresponding to these to be $\frac{3}{8}$, $\frac{35}{8}$, $\frac{63}{8}$, $\frac{99}{8}$, etc. (Proposition 175), and these numbers will occupy the even spaces in row 3 of Table III.

Similarly it is shown (Propositions 176 and 177) that the pyramidal numbers in row four can be expressed by the general formula

$$\frac{n^3+3n^2+2n}{6} \text{ or } \frac{n(n+1)(n+2)}{1\times 2\times 3},$$

where n is the index of the power indicated at the head of the table. This being so, substitution of the values $n=\frac{1}{2}, \frac{3}{2}, \frac{5}{2}, \frac{7}{2}$ gives rise to the numbers

$$\frac{15}{48}, \frac{105}{48}, \frac{315}{48}, \frac{693}{48}, \frac{1287}{48},$$

which accordingly fall into the even spaces of row four (Pyramidales).

* "Propositum sit inquirere, quam habeant rationem numeri Triangulares, ad suum latus." (Prop. clxxi.)

TABLE IV.

		1		n		$\frac{n(n+1)}{1\cdot 2}$		$\frac{n(n+1)(n+2)}{1\cdot 2\cdot 3}$		$\frac{n(n+1)(n+2)(n+3)}{1\cdot 2\cdot 3\cdot 4}$
Row 1	8	1	$\frac{1}{2}$		$\frac{3}{8}$		$\frac{15}{48}$		$\frac{105}{384}$	
Row 2. Monadici	1	1	1	1	1	1	1	1	1	1
Row 3		1	□	$\frac{3}{2}$		$\frac{15}{8}$		$\frac{105}{48}$		$\frac{945}{384}$
Row 4. Laterales	$\frac{1}{2}$	1	$\frac{3}{2}$	2	$\frac{5}{2}$	3	$\frac{7}{2}$	4	$\frac{9}{2}$	5
Row 5		1		$\frac{5}{2}$		$\frac{35}{8}$		$\frac{315}{48}$		$\frac{3465}{384}$
Row 6. Triangulares	$\frac{3}{8}$	1	$\frac{15}{8}$	3	$\frac{35}{8}$	6	$\frac{63}{8}$	10	$\frac{99}{8}$	15
Row 7		1		$\frac{7}{2}$		$\frac{63}{8}$		$\frac{693}{48}$		$\frac{9009}{384}$
Row 8. Pyramidales	$\frac{15}{48}$	1	$\frac{105}{48}$	4	$\frac{315}{48}$	10	$\frac{693}{48}$	20	$\frac{1287}{48}$	35
Row 9		1		$\frac{9}{2}$		$\frac{99}{8}$		$\frac{1287}{48}$		$\frac{19305}{384}$
Row 10. Triangulipyram.	$\frac{105}{384}$	1	$\frac{945}{384}$	5	$\frac{3465}{384}$	15	$\frac{9009}{384}$	35	$\frac{19305}{384}$	70

(Right-hand column labels: Monadici, Laterales, Triangulares, Pyramidales, Trianguli-Pyram.)

CHAPTER IV

And finally, by showing that the terms of the fifth row (Triangulipyramidales) are expressible by the general formula (Proposition 179)

$$\frac{n^4+6n^3+11n^2+6n}{24}, \text{ or } \frac{n(n+1)(n+2)(n+3)}{1\times 2\times 3\times 4\times},$$

and by substituting the values $\frac{1}{2}, \frac{3}{2}, \frac{5}{2}, \frac{7}{2}$ for n, the numbers which are to fall in the even spaces of this row are shown to be

$$\frac{105}{384}, \frac{945}{384}, \frac{3465}{384}, \frac{9009}{384}, \frac{19305}{384}, \frac{36465}{384}.$$

Remembering that the columns merely repeat the rows many of the gaps in Table III can be filled, and thus we arrive at Table IV.

But as Wallis points out in the Scholium to Proposition 185 the new series (row 3) which has arisen through the interpolation of □ is far from complete. The completion of this row and of the whole table constitutes the next step in his elaborate investigation. By an amazingly careful and patient analysis he observes that in any individual series of Table IV, if the first term is denoted by A and the second term (*i. e.*, the first even term) by 1, all the remaining terms of the series (even as well as odd) are obtained by the continued multiplication of the following numbers, namely * :

	Odd.		Even.	
First Row	$A\times \frac{0\times 2\times 4\times 6}{1\times 3\times 5\times 7}$	etc.	$1\times \frac{1\times 3\times 5\times 7}{2\times 4\times 6\times 8}$	etc.
Second Row	$A\times \frac{1\times 3\times 5\times 7}{1\times 3\times 5\times 7}$	,,	$1\times \frac{2\times 4\times 6\times 8}{2\times 4\times 6\times 8}$	,,
Third Row	$A\times \frac{2\times 4\times 6\times 8}{1\times 3\times 5\times 7}$	,,	$1\times \frac{3\times 5\times 7\times 9}{2\times 4\times 6\times 8}$	,,
Fourth Row	$A\times \frac{3\times 5\times 7\times 9}{1\times 3\times 5\times 7}$	,,	$1\times \frac{4\times 6\times 8\times 10}{2\times 4\times 6\times 8}$	,,
Fifth Row	$A\times \frac{4\times 6\times 8\times 10}{1\times 3\times 5\times 7}$	,,	$1\times \frac{5\times 7\times 9\times 11}{2\times 4\times 6\times 8}$	,,
Sixth Row	$A\times \frac{5\times 7\times 9\times 11}{1\times 3\times 5\times 7}$	,,	$1\times \frac{6\times 8\times 10\times 12}{2\times 4\times 6\times 8}$	,,
Seventh Row ...	$A\times \frac{6\times 8\times 10\times 12}{1\times 3\times 5\times 7}$	,,	$1\times \frac{7\times 9\times 11\times 13}{2\times 4\times 6\times 8}$	,,

To put it another way, if in any row, the first two terms are A and 1 respectively, all the terms of that row are :

* This important step, which is the corner-stone of the whole investigation, is enunciated in Prop. 188. "In singulis seriebus Tabellæ Prop. 184. Si primus terminus dicatur A, secundus (hoc est, primus parium,) 1. Reliqui omnes ejusdem seriei (tam pares quam impares) fiunt continua multiplicatione numerorum sequentium."

Row.	Odd.	Even.	Odd.	Even.	Odd.	Even.	Odd.	Even.	Odd.	Even.
1st	A	1	$A \times \dfrac{0}{1}$	$1 \times \dfrac{1}{2}$	$A \times \dfrac{0 \times 2}{1 \times 3}$	$1 \times \dfrac{1 \times 3}{2 \times 4}$	$A \times \dfrac{0 \times 2 \times 4}{1 \times 3 \times 5}$	$1 \times \dfrac{1 \times 3 \times 5}{2 \times 4 \times 6}$	$A \times \dfrac{0 \times 2 \times 4 \times 6}{1 \times 3 \times 5 \times 7}$	$1 \times \dfrac{1 \times 3 \times 5 \times 7}{2 \times 4 \times 6 \times 8}$
2nd	A	1	$A \times \dfrac{1}{1}$	$1 \times \dfrac{2}{2}$	$A \times \dfrac{1 \times 3}{1 \times 3}$	$1 \times \dfrac{2 \times 4}{2 \times 4}$	$A \times \dfrac{1 \times 3 \times 5}{1 \times 3 \times 5}$	$1 \times \dfrac{2 \times 4 \times 6}{2 \times 4 \times 6}$	$A \times \dfrac{1 \times 3 \times 5 \times 7}{1 \times 3 \times 5 \times 7}$	$1 \times \dfrac{2 \times 4 \times 6 \times 8}{2 \times 4 \times 6 \times 8}$
3rd	A	1	$A \times \dfrac{2}{1}$	$1 \times \dfrac{3}{2}$	$A \times \dfrac{2 \times 4}{1 \times 3}$	$1 \times \dfrac{3 \times 5}{2 \times 4}$	$A \times \dfrac{2 \times 4 \times 6}{1 \times 3 \times 5}$	$1 \times \dfrac{3 \times 5 \times 7}{2 \times 4 \times 6}$	$A \times \dfrac{2 \times 4 \times 6 \times 8}{1 \times 3 \times 5 \times 7}$	$1 \times \dfrac{3 \times 5 \times 7 \times 9}{2 \times 4 \times 6 \times 8}$
4th	A	1	$A \times \dfrac{3}{1}$	$1 \times \dfrac{4}{2}$	$A \times \dfrac{3 \times 5}{1 \times 3}$	$1 \times \dfrac{4 \times 6}{2 \times 4}$	$A \times \dfrac{3 \times 5 \times 7}{1 \times 3 \times 5}$	$1 \times \dfrac{4 \times 6 \times 8}{2 \times 4 \times 6}$	$A \times \dfrac{3 \times 5 \times 7 \times 9}{1 \times 3 \times 5 \times 7}$	$1 \times \dfrac{4 \times 6 \times 8 \times 10}{2 \times 4 \times 6 \times 8}$
5th	A	1	$A \times \dfrac{4}{1}$	$1 \times \dfrac{5}{2}$	$A \times \dfrac{4 \times 6}{1 \times 3}$	$1 \times \dfrac{5 \times 7}{2 \times 4}$	$A \times \dfrac{4 \times 6 \times 8}{1 \times 3 \times 5}$	$1 \times \dfrac{5 \times 7 \times 9}{2 \times 4 \times 6}$	$A \times \dfrac{4 \times 6 \times 8 \times 10}{1 \times 3 \times 5 \times 7}$	$1 \times \dfrac{5 \times 7 \times 9 \times 11}{2 \times 4 \times 6 \times 8}$
6th	A	1	$A \times \dfrac{5}{1}$	$1 \times \dfrac{6}{2}$	$A \times \dfrac{5 \times 7}{1 \times 3}$	$1 \times \dfrac{6 \times 8}{2 \times 4}$	$A \times \dfrac{5 \times 7 \times 9}{1 \times 3 \times 5}$	$1 \times \dfrac{6 \times 8 \times 10}{2 \times 4 \times 6}$	$A \times \dfrac{5 \times 7 \times 9 \times 11}{1 \times 3 \times 5 \times 7}$	$1 \times \dfrac{6 \times 8 \times 10 \times 12}{2 \times 4 \times 6 \times 8}$
7th	A	1	$A \times \dfrac{6}{1}$	$1 \times \dfrac{7}{2}$	$A \times \dfrac{6 \times 8}{1 \times 3}$	$1 \times \dfrac{7 \times 9}{2 \times 4}$	$A \times \dfrac{6 \times 8 \times 10}{1 \times 3 \times 5}$	$1 \times \dfrac{7 \times 9 \times 11}{2 \times 4 \times 6}$	$A \times \dfrac{6 \times 8 \times 10 \times 12}{1 \times 3 \times 5 \times 7}$	$1 \times \dfrac{7 \times 9 \times 11 \times 13}{2 \times 4 \times 6 \times 8}$

CHAPTER IV

TABLE V. (Prop. 189.)

	Col. 1	Col. 2	Col. 3	Col. 4	Col. 5	Col. 6	Col. 7	Col. 8	Col. 9	Col. 10	A.
Row 1	8	1	$\frac{1}{2}\square$	$\frac{1}{2}$	$\frac{1}{3}\square$	$\frac{3}{8}$	$\frac{4}{15}\square$	$\frac{15}{48}$	$\frac{8}{35}\square$	$\frac{105}{384}$	1
Row 2	1	1	1	1	1	1	1	1	1	1	
Row 3	$\frac{1}{2}\square$	1	$\square$	$\frac{3}{2}$	$\frac{4}{3}\square$	$\frac{15}{8}$	$\frac{8}{5}\square$	$\frac{105}{48}$	$\frac{64}{35}\square$	$\frac{945}{384}$	$A \times \frac{2 \times 4}{1 \times 3}\cdots\,;\ 1 \times \frac{3 \times 5}{2 \times 4}\cdots$
Row 4	$\frac{1}{2}$	1	$\frac{3}{2}$	2	$\frac{5}{2}$	3	$\frac{7}{2}$	4	$\frac{9}{2}$	5	$A \times \frac{3 \times 5}{1 \times 3}\cdots\,;\ 1 \times \frac{4 \times 6}{2 \times 4}\cdots$
Row 5	$\frac{1}{3}\square$	1	$\frac{4}{3}\square$	$\frac{5}{2}$	$\frac{8}{3}\square$	$\frac{35}{8}$	$\frac{64}{15}\square$	$\frac{315}{48}$	$\frac{128}{21}\square$	$\frac{3465}{384}$	$A \times \frac{4 \times 6}{1 \times 3}\cdots\,;\ 1 \times \frac{5 \times 7}{2 \times 4}\cdots$
Row 6	$\frac{3}{8}$	1	$\frac{4}{3}\square$	3	$\frac{35}{8}$	6	$\frac{63}{8}$	10	$\frac{99}{8}$	15	$A \times \frac{5 \times 7}{1 \times 3}\cdots\,;\ 1 \times \frac{6 \times 8}{2 \times 4}\cdots$
Row 7	$\frac{4}{15}\square$	1	$\frac{8}{5}\square$	$\frac{7}{2}$	$\frac{64}{15}\square$	$\frac{63}{8}$	$\frac{128}{15}\square$	$\frac{693}{48}$	$\frac{512}{35}\square$	$\frac{9009}{384}$	$A \times \frac{6 \times 8}{1 \times 3}\cdots\,;\ 1 \times \frac{7 \times 9}{2 \times 4}\cdots$
Row 8	$\frac{15}{48}$	1	$\frac{105}{48}$	4	$\frac{315}{48}$	10	$\frac{693}{48}$	20	$\frac{1287}{48}$	35	$A \times \frac{7 \times 9}{1 \times 3}\cdots\,;\ 1 \times \frac{8 \times 10}{2 \times 4}\cdots$
Row 9	$\frac{8}{35}\square$	1	$\frac{64}{35}\square$	$\frac{9}{2}$	$\frac{128}{21}\square$	$\frac{99}{8}$	$\frac{512}{35}\square$	$\frac{1287}{48}$	$\frac{1024}{35}\square$	$\frac{19305}{384}$	$A \times \frac{8 \times 10}{1 \times 3}\cdots\,;\ 1 \times \frac{9 \times 11}{2 \times 4}\cdots$
Row 10	$\frac{105}{384}$	1	$\frac{945}{384}$	5	$\frac{3465}{384}$	15	$\frac{9009}{384}$	35	$\frac{19305}{384}$	70	$A \times \frac{9 \times 11}{1 \times 3}\cdots\,;\ 1 \times \frac{10 \times 12}{2 \times 4}\cdots$

In Proposition 189 Wallis shows how, by the application of this rule, all the gaps in Table IV can be filled. It will be at once noted, however, that the rule reduces all the odd terms in the first row to zero, and before proceeding further Wallis made a desperate attempt to bring these numbers into line with his scheme. This he does by making the first term of that row infinite! And now he says (p. 109), just as $\frac{1}{\infty} = 0$, $\frac{2}{\infty} = 0$, $\frac{3}{\infty} = 0$, may we not just as truly say $\infty \times 0 = 1$, $\infty \times 0 = 2$, $\infty \times 0 = 3$; and this ingenious observation gives him an excuse for filling in the blank spaces with finite quantities whilst yet adhering to his plan. This is enunciated in Proposition 188, in which he declares:

"If anyone is diffident concerning the odd numbers of the first row, which I say are obtained by the continued multiplication of the numbers

$$A \times \frac{0 \times 2 \times 4 \times 6 \times 8,}{1 \times 3 \times 5 \times 7 \times 9}$$

inasmuch as since one term of the product is zero, the whole product will vanish, it must be known that we have anticipated the danger in this way; that the term A in that series should be Infinite, (as we showed above in the Scholia to Prop. 116) and that therefore unless zero followed—in order to diminish the 'magnitude' of Infinity—all the terms of that series would be increased to Infinity"*.

Proceeding on these lines, and remembering that the columns repeat the rows, the table can be completed and, in fact, extended indefinitely. Having completed as far as is shown (Table V, p. 57), the final step is to demonstrate how it can be used to evaluate □ or $\frac{4}{\pi}$, though he had already realized that an exact determination of this quantity was impossible. To effect this evaluation he now examines the completed table, and directs attention to the third

* "Si quis autem hæsitat de imparibus seriei primæ, (quos aio fieri ex continua multiplicatione numerorum $A \times \frac{0 \times 2 \times 4 \times 6 \times 8}{1 \times 3 \times 5 \times 7 \times 9}$ etc.) nempe, ne 0. ciphra quæ istic conspicitur, totam continuam multiplicationem, quantacunque fuerit, penitus destruat, faciatque omnes istius seriei terminos evanescere in 0 ciphram seu Nihil: Sciendum est, inde huic malo cautum esse quod terminus A in ista serie sit ∞ Infinitum (prout superius ostendimus in Scholiis ad prop. 166). adeoque nisi sequeretur 0 (ad ipsius ∞ vires minuendas) excrevissent omnes istius seriei termini in ∞ Infinitum."

CHAPTER IV

row in particular, *i. e.*, the row in which $\square$ was interpolated. This, it will be observed, is

$$\frac{1}{2}\square,\ 1,\ \square,\ \frac{3}{2},\ \frac{4}{3}\square,\ \frac{3\times5}{2\times4},\ \frac{4\times6}{3\times5}\square,\ \frac{3\times5\times7}{2\times4\times6},\ \frac{4\times6\times8}{3\times5\times7}\square.$$

Now suppose we call these
$$\alpha,\ a,\ \beta,\ b,\ \gamma,\ c,\ \delta,\ d,\ \text{etc.}\ ;$$
then it is clear that

$$\frac{\beta}{\alpha} = \frac{2}{1},$$

$$\frac{b}{a} = \frac{3}{2},$$

$$\frac{\gamma}{\beta} = \frac{4}{3},$$

$$\frac{c}{b} = \frac{5}{4},$$

$$\frac{\delta}{\gamma} = \frac{6}{5},\ \text{and so on.}$$

It is noticed that these ratios continually decrease. Moreover, for three consecutive terms, *e. g.*, α, a, β, $\frac{a}{\alpha} > \frac{\beta}{a}$, therefore $a^2 > \alpha\beta$. This being so, it is easily shown that

$1 > \frac{1}{2}\square^2,$ whence $\square < \sqrt{2},$

$\square^2 > \frac{3}{2},$,, $\square > \sqrt{\frac{3}{2}},$

$\frac{3\times 3}{2\times 2} > \frac{4}{3}\square^2,$,, $\square < \frac{3\times 3}{2\times 4}\sqrt{\frac{4}{3}},$

$\frac{4\times 4}{3\times 3}\square^2 > \frac{3\times 3\times 5}{2\times 2\times 4},$,, $\square > \frac{3\times 3}{2\times 4}\sqrt{\frac{5}{4}},$

$\frac{3\times 3\times 5\times 5}{2\times 2\times 4\times 4} > \frac{4\times 4\times 6}{3\times 3\times 5}\square^2,$,, $\square < \frac{3\times 3\times 5\times 5}{2\times 4\times 4\times 6}\sqrt{\frac{6}{5}},$

$\frac{4\times 4\times 6\times 6}{3\times 3\times 5\times 5}\square^2 > \frac{3\times 3\times 5\times 5\times 7}{2\times 2\times 4\times 4\times 6},$,, $\square > \frac{3\times 3\times 5\times 5}{2\times 4\times 4\times 6}\sqrt{\frac{7}{6}}.$

Proceeding in this way, Wallis easily shows that $\square$ is less than

$$\frac{3\times 3\times 5\times 5\times 7\times 7\times 9\times\ 9\ \times 11\times 11\times 13\times 13}{2\times 4\times 4\times 6\times 6\times 8\times 8\times 10\times 10\times 12\times 12\times 14} \times \sqrt{1\frac{1}{13}}$$

and greater than

$$\frac{3\times 3\times 5\times 5\times 7\times 7\times 9\times\ 9\ \times 11\times 11\times 13\times 13}{2\times 4\times 4\times 6\times 6\times 8\times 8\times 10\times 10\times 12\times 12\times 14} \times \sqrt{1\frac{1}{14}}.$$

sults tend to equality (*majoris et minoris differentia
...avis assignata minor*, Prop. 191), whence emerges the

$$\square = \left(\frac{4}{\pi}\right) = \frac{3\times 3\times 5\times 5\times 7\times 7\times 9\times 9 \times 11\times 11\times}{2\times 4\times 4\times 6\times 6\times 8\times 8\times 10\times 10\times 12\times}.$$

Wallis also expressed the result in the form

$$\square = 1\times \frac{9}{8}\times \frac{25}{24}\times \frac{49}{48}\times \frac{81}{80}\times \frac{121}{120}\times,$$

" Or, (calling the first term A, the second B, the third C and so on)

$$1+\frac{1}{8}A+\frac{1}{24}B+\frac{1}{48}C+\frac{1}{80}D,$$

so that by a continued Addition of such an Aliquote part of the number last found, you obtained the number sought to what accurateness you please". (*Algebra*, p. 317.)

Nevertheless Wallis felt that the manner in which he had expressed his result could be still further improved, for instead of a finite number of terms yielding an absolute value it contained merely an infinite number approaching nearer and nearer to the true value. He therefore submitted it to Lord Brouncker. He did not succeed in getting an absolute value, but he gave the following beautiful expression—an expression which gave rise to the theory of continued fractions. According to Brouncker $\square$ is to be expressed by the ratio 1 to

$$\cfrac{1}{1+\cfrac{1}{2+\cfrac{9}{2+\cfrac{25}{2+\cfrac{49}{2+\text{ etc.,}}}}}}$$

continued to infinity. This was an improvement upon Wallis's expression, inasmuch as it approached nearer and nearer the true value, first by being in excess of it, and then by being less than it, according as the number of terms was odd or even.

A rich harvest of discoveries is the usual consequence of the discovery of a new method; this reward was not denied Wallis. His *Arithmetica Infinitorum* contained in its pages a multitude of geometrical novelties; these, however, were but a fraction of the many which resulted from the application of the methods he had taught. The celebrated problems of Pascal, most of which were solved by Wallis, are an instance. In 1659 he gave the measure of the surface of the cissoid and of the conchoid, as well as that of solids formed by their revolution. Following the methods which he had elaborated in this monumental treatise he showed the equality of the length of the parabola and the spiral, and he demonstrated that their rectification depended upon the quadrature of the hyperbola. These and many similar researches formed the subject of his treatises on the cycloid and the cissoid. Perhaps more original even than these were the comprehensive results he obtained in 1670 when he first published his treatise *De Centro Gravitatis*. This work, which seemed to contain all that geometry could contribute to statics, will be described later; we pass meanwhile to a consideration of the influence of the *Arithmetica Infinitorum* upon Wallis's contemporaries and his immediate successors.

The Binomial Theorem was the direct outcome of a consideration of Wallis's treatise. It has already been pointed out that Wallis did not succeed in his interpolation; the very methods which brought him so near his goal likewise convinced him that such interpolation was impossible; for he knew that on expansion the expression $(\sqrt[p]{R} - \sqrt[p]{a})^n$ would have $(n+1)$ terms so long as n was a positive integer. If n were $\frac{1}{2}$, as the quadrature of the circle demanded, the emergent expression should have "more terms than one, and less than two", and even his genius could not see how this *impasse* could be avoided. He concluded that the required value could not be expressed in any known way of notation either rational or irrational, and therefore a new one must be devised to express it.

With the extravagant symbolism of the *Clavis* no doubt in his mind he therefore essayed to invent a new notation to express the quantity sought. "We have already a symbolism", says Wallis, in the Scholium to Proposition 190, "to express quantities which are known to be indeterminate. For example, the symbol $\sqrt{\ } : 3 \times 6$ denotes an indeterminate quantity, since it represents the term intermediate between the first and the second terms of the Geometrical Progression, 3, 6, 12, (numbers obtained by the continued multiplication of $3 \times 2 \times 2 \times 2$ etc.). In the same way, I shall use the symbol $nr : 1 \mid \frac{3}{2}$ to denote the term interpolated between the first and the second of the series $1, \frac{3}{2}, \frac{15}{8}$, etc., (which are obtained by the continued multiplication of the numbers $1, \frac{3}{2}, \frac{5}{4}$, etc.).

Further, the circle is to the square on the diameter as 1 is to $\mathit{nr} : 1 \mid \frac{3}{2}$.
And this indeed will be the true quadrature of the circle " *.

It remained for Newton to complete the investigation. " At the beginning of my mathematical work ", he wrote in a famous letter sent to Oldenburg for transmission to Leibniz, " when I came across the work of Wallis, I considered the series by the intercalation of which he gives the area of the circle and of the hyperbola " †. Newton's statement, frank as it is, hardly however conveys the true extent of his indebtedness to Wallis. The resemblance, for example, between Wallis's results exhibited in the foregoing pages and Newton's opening paragraph of the treatise *De Analysi per Æquationes Numero Terminorum Infinitas*, where he states

$$\text{Si } ax^{m/n} = y,$$

Erit $\dfrac{an}{m+n} \cdot x^{\frac{m+n}{n}} = \text{Area ABD},$ [Fig. 8.]

is too striking to be fortuitous.

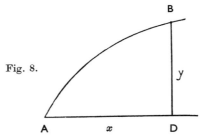

Fig. 8.

Moreover Wallis's methods of quadrature were diligently pursued by his contemporaries. Lord Brouncker obtained the first infinite series for the area of the equilateral hyperbola between

* " Si igitur ut $\sqrt{\ } : 3 \times 6 :$ significat *terminum medium inter* 3 *et* 6 *in progressione Geometrica æquabili* 3, 6, 12, etc., (continue multiplicando $3 \times 2 \times 2$ etc.) ita $\mathit{nr} : 1 \mid \frac{3}{2} :$ significet *terminum medium inter* 1 *et* 3/2 *in progressione Geometrica decrescento* 1, 3/2, 15/8 etc. (continue multiplicando $1 \times 3/2 \times 5/4$ etc.) erit $\square = \mathit{nr} : 1 \mid \frac{3}{2} :$ Et propterea *circulus est ad quadratum diametri, ut* 1 *ad* $\mathit{nr} : 1 \mid \frac{3}{2}.$ Quæ quidem erit vera circuli quadratura in numeris, quatenus ipsa numerorum natura patitur, explicata."

† " Sub initio studiorum meorum Mathematicorum, ubi incideram in opera Celeberrimi Wallisii nostri, considerando Series quarum intercalatione exhibet Aream Circuli, etc." Newton to Oldenburg, Oct. 24, 1676. Wallis, *Opera*, iii, p. 634.

CHAPTER IV

its asymptotes, and it is to the *Arithmetica Infinitorum* that Mercator owed the invention of his well-known series for the area of the hyperbola. It was in seeking to apply the rules which had been elaborated by Wallis to the hyperbola that he evolved the series which expresses the area of the hyperbolic space between its asymptotes. It was a consequence of Wallis's investigation that if the ordinate of a curve could be expressed by a series of powers of the abscissa, as, for example,

$$1+x+x^2+x^3+x^4+,$$

the area under the curve would be

$$x+\frac{x^2}{2}+\frac{x^3}{3}+\frac{x^4}{4}+\frac{x^5}{5}+,$$

Wallis had also observed that, by taking the origin on the asymptote at a distance from the centre equal to BC or unity (fig. 9), so that BD$=x$, the ordinate is $\frac{1}{1+x}$, but this expression did not fall within the scope of the rules he had evolved, and therefore his efforts to accomplish its quadrature were fruitless. But Mercator saw further than Wallis (Mercator, *Logarithmotechnia*, Prop. 7). By the usual process of division he perceived that $\frac{1}{1+x}$ could be written $1-x+x^2-x^3+x^4-$, and so long as x was less than 1 this is equivalent to two decreasing geometrical progressions, namely,

and
$$\begin{aligned}1+x^2+x^4+x^6+\\-x-x^3-x^5-x^7,\end{aligned}$$

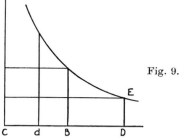

Fig. 9.

and therefore their sum could be readily effected. Equally important was an observation in the Scholium to Proposition 38, in which Wallis came very near the solution of a still more difficult problem, namely, that of Rectification. How near the fringe of the solution he actually approached is perhaps best described in his own words, written thirty years later in his *Algebra* (p. 293). "I was then aware", he says, "that in case a Curve were so ordered that the differences of the Ordinates to the Axe, (whether on the Concave or on the Convex side) taken at equal distances, be as the Quadratick Roots of numbers in Arithmetical Progression, as $\sqrt{0}$, $\sqrt{1}$, $\sqrt{2}$, $\sqrt{3}$, etc. (as are the Ordinates in a Parabola) and therefore their Squares as 0, 1, 2, 3, 4, these Squares increased by the square of the Intervalls of such Ordinates, suppose by 4, (the square of 2) will be the Squares of the Subtenses to the portions of the Curves; as 4, 5, 6, 7, etc. (which in parts infinitely small, are coincident with the Curves;) as are the Ordinates in a Trunk of the same Parabola. And consequently, as a Parabola to a Trunk of the same Parabola, so is the base, (or Aggregate of those differences) to the Curve (or Aggregate of such subtenses)". Wallis contented himself with this observation, but the hint did not prove fruitless. A young geometer, William Neile, went much further. He noticed that in order that the second curve which we want to describe was absolutely quadrable it was necessary that the differences of the ordinates of the first were as the ordinates of a parabola, and that then this new curve which resulted therefrom was a section of a parabola, whence he concluded that the first curve was absolutely rectifiable.

Neither Wren nor Brouncker perceived the nature of this remarkable curve, and it was fitting that Wallis himself, who had taken the first steps in its discovery, should have a further share in it. He recognized that the curve under discussion was one of the cubical parabolas, in which the cube of the ordinate is proportional to the square of the abscissa. Shortly afterwards Wren also discovered the rectification of the cycloid, and thus, contrary to what Descartes had maintained, there were two curves, one geometrical and the other mechanical, which were rectifiable.

"The *Arithmetica Infinitorum* has ever been acknowledged to be the fountain of all the Improvements that have been made in Geometry since that time" *. This was the considered opinion of Doctor Gregory in the *Life of Wallis*, to which reference is made elsewhere, and it is clear from the foregoing pages that the claims he made on his behalf were by no means extravagant. Nevertheless, the treatise met with not a little hostile criticism, and in particular Fermat was quick to detect its blemishes. This will be described in due course. But in spite of this, contemporary mathematicians at once realized what a valuable contribution to science this treatise constituted, and it is probably not too much to say that, despite its shortcomings, few works in the history of science have exerted greater influence or been more universally admired.

* Gregory, *Life of Wallis* (Smith MS. 31). (Bodleian.)

CHAPTER V.

THE *MATHESIS UNIVERSALIS*.
THE *COMMERCIUM EPISTOLICUM* (1657/8).
CONTROVERSY WITH FERMAT AND OTHER MEMBERS OF THE
FRENCH MATHEMATICAL SCHOOL.

THE *Arithmetica Infinitorum* was the first of a long series of mathematical treatises which came from Wallis's pen. These followed in rapid succession, and scarcely a year passed but that the world of learning was enriched with some erudite contribution from him. This amazing vitality is all the more striking when it is recalled that no sooner had Wallis become recognized as one of the foremost mathematicians of the day than he found himself entangled in a maze of controversy, from which he did not completely extricate himself until the closing years of his life. It was in 1656, for example, whilst he was exposing the mathematical ineptitude of Thomas Hobbes, that he published his treatise on the *Angle of Contact*[*], i.e., the angle between the curved line and its tangent.

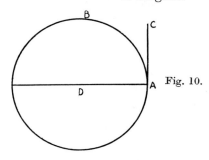

Fig. 10.

It had been demonstrated by Euclid that the line CA (fig. 10), standing perpendicular to a radius DA touches the circumference at one point only, and that no right line could be drawn between the tangent and the circle. Hence the angle of contact is less than any rectilinear angle, and the angle of the semicircle between the radius DA and the arc AB is greater than any rectilinear acute angle. This seeming paradox of Euclid had exercised the best minds of the Middle Ages, and in particular it was the subject of a long controversy between Peletarius and Clavius, the former of whom

* *De Angulo Contactus, et Semicirculi, Disquisitio Geometrica*, Oxon, 1656.

maintained that the angle of contact was heterogeneous to a rectilinear one, as a line to a surface, the latter maintaining the contrary. Wallis lost no time in entering the lists. He promptly espoused the opinions of Peletarius, asserting dogmatically that the " angle of contact was of no magnitude and not any part of a Rightlined Angle " (*Phil. Trans.*, 1695, p. 74). Subsequently he published a work *A Defense of the Treatise of the Angle of Contact*. In this he sharply criticized not only Clavius, but also his brother Jesuit, Leotaud, and he summarized his views in the words : " I do show, (with Peletarius against Clavius) that (what is commonly called) *the Angle of Contact*, is *of no magnitude*. But is, to a real Angle, whether Rectilinear or Curvilinear .... as 0, (a Cypher) to a Number " (Chap. I). Using the ancient Method of Exhaustions, which in turn depends for its validity upon the Euclidean proposition that " those quantities whose difference is less than any assignable quantity are equal ", Wallis argued that the quantity called " *somewhat* " by Clavius, as representing the difference between the two angles made by the tangent and the circle with the diameter at the point of contact, has really no magnitude, or at least is conceived as such in geometry. " No part of any magnitude ", said Wallis, " can be so small but that it may be so multiplied as to equal or exceed the whole ".

In this treatise Wallis introduced the idea of " inceptive quantities ", and his definition of these—" *prima principia quod sic* "— no doubt provided Newton with a hint for the " nascent quantities " which are at the root of the Method of Fluxions. " There are some things ", said Wallis, " which tho' as to some kind of Magnitude, they are nothing ; yet are in the next possibility of being somewhat. They are *not* it, but *tantum non* ; they are in the next possibility to it ; and the Beginning of it : Tho' not as *primum quod sit*, (as the Schools speak) yet as *ultimum quod non*. And may very well be called *Inchoatives* or *Inceptives*, of that somewhat to which they are in such possibility. Thus the point A .... is in the next possibility to Length, and Inceptive of it. (For, if it never so little moved, it describes a Line ") *.

The following year (1657) there appeared his *Mathesis Universalis, sive Opus Arithmeticum*, which was dedicated to Gerard Langbaine, Henry Wilkinson, John Wilkins, and Jonathan Goddard. Indisputably the most valuable feature of this work is its treatment of notation. The development of arithmetical notation from antiquity is traced with a fullness hitherto unknown. Algebraic notation, which had never kept pace with the rapid developments in other branches of algebra during the sixteenth century, now at last begins

* *Defense of the Treatise of the Angle of Contact*, pp. 95/96.

CHAPTER V

to assume a more modern aspect. Wallis gave this development a great impulse. He first of all called attention to the disadvantage of denoting algebraic powers by geometrical dimensions, the employment of which had already led to marked ambiguity. "Algebraical powers are better explained by means of arithmetical degrees than by geometrical dimensions" he declared *. His close acquaintance with the history of algebra is reflected in his familiarity with the different notations which have been current during various periods. He uses these to illustrate the conciseness of the modern, and to emphasize the superiority of the latter he performs many of his operations as many as five times, employing a different notation each time, so that from a comparison of all these the deficiencies of the older forms are at once apparent. These different notations are also exhibited in a table (here reproduced), and his remarks thereon are extremely illuminating :

Nomina.	Stifel.	Vieta.	Oughtred.	Harriot.	Descartes.	Potestas seu gradus.
Radix	℞	R	A	a	a	1
Quadratum	℥	Q	Aq	aa	a^2	2
Cubus	ℭ	C	Ac	aaa	a^3	3
Quad. quadratum	℥℥	QQ	Aqq	$aaaa$	a^4	4
Surdesolidum	ſſ	S	Aqc	&c.	a^5	5
Quad. Cubi	℥ℭ	QC	Acc	...	a^6	6
2 Surdesolidum	Bſſ	bS	Aqqc	...	a^7	7
Quad. quad. quad.	℥℥℥	QQQ	Aqcc	...	a^8	8
Cubi cubus	ℭℭ	CC	Accc	...	a^9	9
Quad. Surdesol.	℥ſſ	QS	Aqqcc	...	a^{10}	10
3ᵐ Surdesolidum	Cſſ	cS	Aqccc	...	a^{11}	11
Quad. quad. cubi	℥℥ℭ	QQC	Acccc	...	a^{12}	12
4ᵐ Surdesolidum	Dſſ	dS	Aqqccc	...	a^{13}	13
Quad. 2 Surdesol.	℥Bſſ	QbS	Aqcccc	...	a^{14}	14
Cubus Surdesol.	ℭſſ	CS	Acccccc	...	a^{15}	15
Quad. quad. quad. quad.	℥℥℥℥	QQQQ	Aqqcccc	...	a^{16}	16
&c.						

In the first column of symbols—"characteres cossici" as Wallis calls them—are the German symbols as found in Stifel, which, he says, had their origin in the letters r, z, c, s, the initial letters of the words *res, zensus, cubus, sursolidus*. In the second column are the letters, R, Q, C, S (radix, quadratum, etc.) and their combinations, which, he notes, were employed by Vieta. In the third column are Oughtred's abbreviations of Vieta's symbols. The fourth column gives Harriot's notation, and in the fifth is the exponential symbolism adopted by Descartes. Apart from this there is no reference to the improvements in notation introduced

* " Potestates Algebraicæ melius per Gradus Arithmeticos, quam per Geometricas Dimensiones, explicantur." *Mathesis Universalis* (Operum Mathematicorum Pars Prima (1657), p. 73.

by the English writers, which, in view of what he wrote thirty years later in his *Algebra*, is very striking. But what is perhaps even more striking is his almost flattering reference to Descartes. " Descartes, and others after him ", he says, " in order to avoid the monotony of frequently repeated letters as formerly, denote the ' root ' by a certain letter of the alphabet, and they indicate the other powers by raised numbers (according to the degree of the power) as a, a^2, a^3, a^4, ... an example of which I have submitted for perusal in the accompanying Table " *.

As a mathematical treatise, however, the *Mathesis Universalis* is a much more elementary work than Wallis was accustomed to deal with, and it has been suggested that it embodied the substance of his professorial lectures. If that be so, it is a striking commentary upon the poverty of the mathematical learning at Oxford of which Wallis had more than once complained. In the early chapters the author discourses at considerable length upon the fundamental operations of arithmetic, but, except that the notation is more understandable and the treatment somewhat fuller, there is nothing here which is not in the first four chapters of Oughtred's *Clavis* or even earlier writers. His description of negative quantities is highly interesting, and shows that even as late as this the idea of negatives still presented difficulties to many a scholar. And it does not seem the least likely that Wallis's attempts at removing the obscurity were at all successful, since he repeatedly calls such quantities *imaginaries*. Why he should do so is not at all clear. The name *negative* appears regularly in his earlier work, the *Arithmetica Infinitorum*, and even in later chapters of the present treatise.

" Impossibile est ut ex 5 subducantur 8, cum numerus hic sit illo major ",—it is impossible to subtract 8 from 5, since the former is greater than the latter. " Yet ", continued Wallis, " although this is impossible, mathematicians and especially algebraists, look upon it as though it were not impossible. For they suppose, besides real quantities, certain *imaginary* quantities which are less than nothing ('*quantitates quasdam imaginarias, quæ minores sint quam nihil, quas quantitates Negativas, sive Ablativas appellant*'), and thus where there is need of distinction, they indicate positive quantities by the sign $+$, and negative or imaginary or ablative by the sign $-$.

" So that if from 5 is to be taken 8, they say -3 remains, that is, that quantity which is the third number less than 0.

* " D. *des Cartes*, et post illum alii, literarum sæpe iterandarum tædium timentes, radicem, ut prius, qualibet Alphabeti litera designant, et ipsius reliquas potestates suspensis notis numericis (pro numero Gradus seu Potestatis) designant, ut, a, a^2, a^3, a^4 etc. Quorum specimen in subjuncto Schemate intuendum proposui," *op. cit.*, p. 71.

CHAPTER V

"Nor is this supposition absurd. For when they say, $5-8=-3$, it is as though they said : He who supposes 8 to be subtracted from 5 supposes a certain third number less than 0. (*Supponit ille aliquid ternario numero minus quam nihil*) " *.

In the early chapters he discourses upon Decimal Fractions, which by this time had become well established. Though in the *Algebra*, written nearly thirty years later, he writes his decimals in the modern form, in the present treatise he follows the clumsy notation adopted by Oughtred. For example, on page 65 he writes the decimal 3579·753 as 3579/753, and he explains the notation thus :

"7, septem partes decimas
5, quinque partes centisimas
3, tres partes millesimas."

The reversion to the notation of Oughtred was unfortunate, and was probably instrumental in delaying the adoption of the comma (or full stop) for many years. Napier, it may be recalled, had used the comma at the beginning of the century (*Rabdologiæ seu numerationis per virgulas libri duo*, Edinburgh, 1617). All this is much inferior to the treatment of decimals which we find in the *Algebra* (chapter lxxxix), where the properties of such numbers are exhaustively treated, and where the nature of circulating decimals is for the first time seriously investigated.

The conception of positive and negative indices which he had developed in the *Arithmetica Infinitorum* is still further explained in chapter xviii. In explaining the above decimal, he writes it

$$\frac{3 \mid 2 \mid 1 \mid 0 \mid \bar{1} \mid \bar{2} \mid \bar{3}}{3 \mid 5 \mid 7 \mid 9 \mid 7 \mid 5 \mid 3},$$

and he adds : " The units place has the index 0 . . . . places above that ascending, have positive indices first 1, second 2, third 3, ... whilst on the other hand, those descending have negative indices, as first -1, second -2, third -3 etc." †. He also explains the sexagesimal notation, and he extends it to fractions. The idea of positive and negative exponents is shown to persist in the sexagesimal notation. " These are first *Minutes*, second, third, etc., as we descend to the right, whilst ascending to the left are units which are called *Sexagena*. First (powers), second, third, etc., are indicated thus—

$$\overset{\backslash\backslash\backslash\backslash}{49}, \overset{\backslash\backslash\backslash}{36}, \overset{\backslash\backslash}{25}, \overset{\backslash}{15}, \overset{\circ}{1}, \overset{\prime}{15}, \overset{\prime\prime}{25}, \overset{\prime\prime\prime}{36}, \overset{\prime\prime\prime\prime}{49},$$

* *Op. cit.*, cap. xiv, pp. 99/100.

† " Unitatis locus indicem habet 0 . . . . loci autem supra illum ascendentes primus, secundus, tertius, etc. indices habent affirmativos 1, 2, 3, et contra loci descendentes primus, secundus, tertius indices habent negativos $-1, -2, -3$," *op. cit.*, cap. xviii.

where any place has a value sixty times greater than the one on its immediate right, and on the other hand, sixty times less than the one on its immediate left " *.

In chapter xxiii are demonstrated algebraically the theorems which comprise the Second Book of Euclid's Elements, and this leads to some important developments in the mensuration of plane figures.

Starting at chapter xxvi the Progressions are investigated. The notation is somewhat crude, and the influence of Oughtred's alleged improvements is plainly visible. In this chapter he proves most of the properties of the Progressions which are to be found in a modern text-book. In particular is his treatment of Geometrical Progressions very comprehensive. The sum of such a progression is usually expressed by him in terms of its first term A, its last term V, and its common ratio R. " Si terminus primus seu minimus dicatur A, maximus V, communis ratio R, et progressionis summa S, erit

$$S = \frac{VR - A}{R - 1}.$$"

Although the rules for the sum of an Arithmetical and a Geometrical Progression had been given long before, notably by Clavius, yet the general problem, namely, given any three of the four quantities, first term, last term, common ratio and the sum, to find the fourth first appears in chapter xxxiii of the *Mathesis Universalis*. Harmonical Progressions are but lightly touched upon. " Tres quantitates Harmonice proportionales sunt, si ut prima ad tertiam, sic differentia primæ et secundæ, ad differentiam secundæ et tertiæ. Puta A . C : : B—A . C—B, erunt, est hoc casu tres A, B, C, Harmonice proportionales. Sic numeri, 3, 4, 6, vel 3, 4, 4, 6. Nam 3 . 6 : : 4—3=1 . 6—4=2." This, of course, accords with the modern definition of such a progression. But the series which precedes it forms no harmonical progression as we know it. " Puta si A . D : : B—A . D—C. Erunt illæ quantitates A, B, C, D, harmonice proportionales. Sic numeri 5, 8, 12, 30, quia 5 . 30 : : 8—5=3 . 30—12=18 " †, that is, four quantities A, B, C, D are harmonically proportional if the first is to the fourth as the difference of the first and second is to the difference of the third and fourth, as 5, 8, 12, 30, since $5 : 30 = 8 - 5 : 30 - 12$. The properties of the arithmetical and the geometrical progressions are concisely summarized in chapter xxviii, but this part of the work, though amazingly comprehensive, and containing some of

* *Op. cit.*, cap. xl, p. 68.
† *Op. cit.*, p. 230.

CHAPTER V

the most erudite portions of the whole work, suffers by an unaccountable relapse into the excessive symbolism of Oughtred. For example, he writes (p. 240) his arithmetical progression thus :—

Puta A, M, V ∸ quoniam est M−A=V−M=E.

Later (chapter xxxvi, *Elementi Quinti Synopsis*) there is a further relapse. Here we meet the following odd combination of symbols :

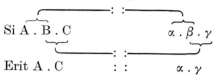

Erit A . C : : α . γ

for what in modern terminology would be written :

If A : B =α : β,
and B : C =β : γ,
then A : C =α : γ.

The main interest of this treatise lies perhaps in the fact that it reflects the state of many of the offshoots of the main mathematical stream during the latter half of the seventeenth century. It probably did little to enhance Wallis's reputation as a mathematician, and it is difficult to believe that the lamp of mathematical knowledge burned so low as it indicates. Nevertheless it evoked a deluge of recriminations from Hobbes, who vigorously attacked it in the pages of his *Examinatio et Emendatio Mathematicæ Hodierniæ* (Lond., 1660). This will be described in due course ; meanwhile a controversy which broke out about this time with the French mathematicians claims attention.

In the spring of 1656/7 Lord Brouncker wrote to Wallis, enclosing the following :—

" A Challenge from M. Fermat, for Doctor Wallis, With the hearty commendations of the messenger, Thomas White : Proponatur (si placet) *Wallisio*, et reliquis Angliæ Mathematicis, sequens Quæstio numerica.

" Invenire Cubum, qui additus omnibus suis partibus aliquotis conficiat Quadratum. Exempli gratia. Numerus 343 est Cubus, à latere 7. Omnes ipsius partes aliquotæ sunt 1, 7, 49 ; quæ adjunctæ ipsi 343, conficiunt numerum 400, qui est quadratus à latere 20. Quæritur alius cubus numerus ejusdem naturæ.

" Quæritur etiam numerus Quadratus qui additus omnibus suis partibus aliquotis conficiet numerum Cubum.

" Has solutiones expectamus ; quas si Anglia aut Galliæ Belgica et Celtica non dederint, dabit Gallia Narbonensis, easque in pignus nascentis amicitiæ D. *Digby* offeret et dicabit " *.

* *Commercium Epistolicum*, i, March 5, 1656/7.

"Let the following numerical question be proposed to Wallis and the other English mathematicians :

"To find a Cube, which added to all its aliquot parts, will make a Square. Example. The number 343 is the cube of 7 ; all its aliquot parts are 1, 7, 49, and these added to the number itself make 400, which is the square of 20. Another number of the same type is required.

"A square number also is required which added to all its aliquot parts will make a cube.

"We await these solutions, which, if England or Belgic or Celtic Gaul do not produce, Narbonese Gaul will, and as a pledge of our growing friendship, Mr. Digby will present and dedicate them."

Wallis seemed disposed to underrate the ability of Fermat, for two days later he replied to Brouncker suggesting that the number 1 satisfied both problems ("Unum eundemque numerum 1, utrique quæsito satisfacere" *). Brouncker, however, was under no delusions regarding the genius of the French mathematician. " I believe the following propositions to be more difficult than at first sight appears", he wrote in the letter which accompanied the problems, " otherwise he scarcely merits the title which he has borne ". Wallis's reply could hardly be expected to satisfy Fermat, who later in the year published a treatise *Solutio duorum Problematum circa Numeros Cubos et Quadratos, quæ tanquam insolubilia universis Europæ Mathematicis*, a Clarissimo Viro D. Fermat, sunt proposita et cet. (a Domino B. Frénicle de Bessy inventa).

This was but the beginning of another protracted quarrel between Wallis and the French mathematical school. Wallis does not appear to have welcomed the type of mathematical joust which was becoming fashionable throughout Europe. " But I added withal ", he wrote to Sir Kenelm Digby, " that I looked upon problems of this nature, (of which it it is easy to contrive a great many in a little time) to have more in them of labour than either of Use or Difficulty " †, and later, when he continued to be harassed by like problems, he complained with some irritation " 'Tis time now that they gave mee leave to mind my own busyness " ‡. But in spite of his spirited rebukes he did not hesitate to reply to his antagonist with a kindred problem :

" Invenire duos numeros quadratos, qui partibus suis aliquotis additi, eandem efficiant summam " §. (To find two square numbers

---

* *Commercium Epistolicum*, ii, 7 Mar., 1656/7.
† *Ibid.*, vii, Sept. 3, 1657.
‡ Wallis to Oldenburg, Nov. 3, 1670. (*R. S. Guard Bks.*)
§ *Commercium Epistolicum*, ii, Mar. 7, 1656/7.

CHAPTER V

which added to their aliquot parts shall make the same sum), *e. g.*,

$$16+8+4+2+1=31=25+5+1.$$

This, however, was sent " not as a new difficulty, but as a trial whether Monsieur *Fermat* did thoroughly understand the mystery of his own two Questions ; and did not only by chance light on them : For if he thoroughly understood those, he must needs be able to solve this with much ease; which it seems by Epist. 37, he did not find so easie ; and therefore, what solution he did find, he chose rather to conceal than let us know it. Nor doth anywhere let us know, whether he were able to solve his own Questions. But Monsieur *Frenicle*, gives solutions both of this and those ; but without acquainting us by what methods he came at them ; which makes me think they are not better than mine " *.

Such a suggestion may not have been entirely unprovoked ; nevertheless Wallis in making it erred greatly. For Fermat was virtually the creator of this branch of mathematics—namely, the Theory of Numbers—and contemporary as well as later mathematicians readily recognized his extraordinary brilliance therein. Even Digby, writing to Wallis on this subject later in the year, admitted that Fermat was " incredibly quick and smart in any thing he taketh in hand " †. Moreover, if Cantor is right, Fermat was the greatest French mathematician of the seventeenth century, notwithstanding the fact that this century had already produced a Descartes and a Pascal ‡. Consequently Wallis, in appraising at such a low estimate Fermat's abilities in this direction, showed a lack of judgment which it is not easy to understand. Nevertheless it was of long standing. In his *Algebra*, published in 1685, when Fermat had been dead nearly twenty years, he wrote : " What were the Methods of *M. Fermat* or *M. Frenicle* herein, I cannot tell : For though they sent us many challenges, (which were performed by us), yet they would never be so kind, (though sometimes they seemed to promise it), as to let us know how themselves performed any of those Problems which they proposed to us ; (save only a lame account, in *M. Frenicle's* Book on this occasion, of some little of what is here perfectly delivered; and that after it had been here done much better). But I think we may well be confident (from their manner of managing these contests,) that if they had better Methods than those of ours, they would have gloried in out-doing us therein. But when they saw that we had, without

---

* *A Discourse on Combinations, Alternations and Aliquot Parts* (1685), p. 151.
† *Commercium Epistolicum*, xxi, Digby to Wallis, Paris, 6 Feb., 1658.
‡ *Bulletin of the American Mathematical Society*, vii (1901), p. 160.

their help, found Methods of our own, as good or better than theirs, they thought it fit to conceal their own " *.

Other problems were speedily forthcoming from Fermat, e.g., " Given any non-quadrate number, there is an infinite number of squares which multiplied into it, and increased by 1, make a square, e.g., 3, (a non-quadrate number) multiplied by 4^2 and increased by 1 gives 49 or 7^2. Required a general rule by which squares of this kind can be found " †.

To this Wallis and Brouncker sent independent solutions ‡.

In the August of 1657, however, Fermat wrote to Digby declaring that the letters he had received contained no solutions, whereupon he proposed others. " Pour les questions des nombres ", he declared, " j'ose Vous dire avec respect et sans rien abbattre de la haute opinion, que j'ay de vostre Nation, que les deux lettres de My Lord *Brouncker*, quoy qu' obscures à mon esgard et mal traduites, n'en contiennent point aucune solution " §. This, together with a casual remark which he threw out, was enough to put Wallis on his mettle. " Ce n'est pas que je pretende par là renouveller les joustes et les anciens coups de lances, que les Anglois ont autrefois fait contre les Francois " ||, wrote Fermat. To all his problems Wallis now replied, and his solutions to them are of such a nature that the merest perusal of them gives incontrovertible evidence of his amazing patience and perspicuity. Indeed, the one outstanding fact which emerges from a consideration of this contest is that Wallis himself had acquired such skill in this branch of mathematics, namely, the Theory of Numbers, as to rank second only to Fermat himself, and there can be little doubt that his genius for problems of this type has never received due recognition. His achievements in this field, brilliant though they were, suffered eclipse by his more fruitful investigations upon the subject of Indivisibles. Nevertheless, no estimate of his mathematical ability can afford to lose sight of his researches into the Theory of Numbers, and it is to be regretted that, in spite of his well-directed efforts to stimulate interest in it, the subject was allowed to fall into desuetude until it was resuscitated by Euler and the Bernouillis after the lapse of nearly a century.

When at length the discourse had run its course Wallis was acknowledged one of the foremost mathematicians in Europe. No better evidence of this can be adduced than two letters which Digby sent from Paris in May 1658. In the first of these, which

* *Algebra*, p. 371.
† *Commercium Epistolicum*, ix, Wallis to Digby, Sept. 27, 1657.
‡ *Ibid*.
§ *Commercium Epistolicum*, xii, Fermat to Digby, Aug. 15, 1657.
|| *Ibid*.

was addressed to Thomas White, he observed: " Truly these last Letters from his Lordship and the Doctor have wrought a mighty change in mens opinions of them. They are now looked upon as the greatest Mathematicians of the Age : And let me tell you this in particular, I asked M: *Frenicle* how he weighed .... in the scale against either of these, he presently replyed that there was no weighing of them together ; for he was but as a slight Scholar in respect of them, the greatest Masters of the Age " *. To Wallis he wrote : " You have now shewed our Mathematicians here, that like *Sampsons* you can easily break and snap asunder all the Philistians cords and snares, when the assault cometh warmly upon you. And the greatest men here, are now forced to allow, that *England* yeildeth to no part of the world in these noble speculations. Mons. *Frenicle* now speaketh loud and high, how much he reverenceth your deep knowledge " †.

Yet when Frénicle suggested publishing the correspondence the idea filled both Brouncker and Wallis with apprehension. That no little share in the honours of the contest rested with Frénicle is established by the merest perusal of the letters, even though Brouncker had petulantly declared in a letter to Wallis that " his Arguments are so inconsiderable, that they hardly deserve any reply " ‡. Nor could Wallis ever bring himself to trust the French mathematicians. " As to our correspondence with the French ", he wrote later to Collins, " I like it best when it is done in print ; they being apt to be disingenuous in claiming all for their own which they have from hence, without owning whence they have it " §. Nevertheless, whether inspired by a genuine desire to focus attention upon the important Theory of Numbers, or, what was much more likely, to exalt the English mathematical school, Wallis wrote to Sir Kenelm Digby in March 1657/8, through whose hands the correspondence had passed, seeking permission to publish the letters. This was readily granted, and the correspondence appeared in 1658 under the title of *Commercium Epistolicum de Quæstionibus quibusdam Mathematicis, nuper habitum.*

But as soon as it appeared it became plain that the dispute between Wallis and Fermat had not been confined to the challenging problems already mentioned. Fermat, it seemed, had made some unhandsome reflections upon the *Arithmetica Infinitorum*, which he criticized, first on account of the excessive amount of symbolism which Wallis had employed in its pages, and secondly on account of the Method of Induction which Wallis had introduced into his

* *Commercium Epistolicum*, xli, May 8, 1658.
† *Commercium Epistolicum*, xlii, May 8, 1658.
‡ *Commercium Epistolicum*, xx, Feb. 18, 1657/8.
§ Wallis to Collins, Sept. 16, 1676 (Rigaud, ii, p. 600).

treatise. The first tremors, apparently, had been heard as early as April 1657, when Fermat wrote to Digby suggesting that the quadratures of the parabolas and the hyperbolas which Wallis had discussed in the *Arithmetica Infinitorum* had already been performed by him many years earlier. " J'ay leu l'*Arithmetica Infinitorum* de *Wallisius*", he averred, " et j'en estime beaucoup l'Autheur. Et bien que la quadrature tant des paraboles, que des hyperboles infinies ait esté faicte par moy depuis fort longues années, et que j'en aye autres fois entretenu l'illustre Toricelli, je ne laisse pas d'estimer l'invention de *Wallisius*, qui sans doubte n'a pas sceu, que j'eusse preoccupé son travail " *. Not unexpectedly the patronizing manner affected by Fermat incensed Wallis beyond measure, and he promptly declared that Fermat's parabolas were but the same as those of which he had already given the quadratures in Propositions 102 and 103 of his own treatise. But the full fury of the tempest broke when Fermat sent to Digby his " Remarques de M. de Fermat sur l'*Arithmetique des Infinis de M. Wallis* ", in which he disparaged Wallis's treatise in no unmeasured terms.

In the Dedication of the *Arithmetica Infinitorum* Wallis had explained how he had reduced one of his main problems to finding a mean between the first and second terms of the series

$$1, \ 6, \ 30, \ 140, \ 630.$$

" In the series of numbers 1, 6, 30, 140, 630, a middle term is sought to be interpolated between 1 and 6. I have indicated that these terms are to be obtained by the continued multiplication of the numbers

$$1 \times \left(4+\frac{2}{1}\right) \times \left(4+\frac{2}{2}\right) \times \left(4+\frac{2}{3}\right) \ldots \text{ or } 1 \times \frac{6}{1} \times \frac{10}{2} \times \frac{14}{3} \times,$$

of which the numerators as well as the denominators are arithmetically proportional " †.

* " I have read the *Arithmetica Infinitorum* of Wallis, and I greatly esteem the author for it. And although the quadrature of many of the parabolas and infinite hyperbolas was done many years before by me, and although I have on many occasions discussed them with the illustrious Torricelli, nevertheless I cannot but admire Wallis's invention, for he could not have known that I had anticipated his work." *Commercium Epistolicum,* iv, Fermat to Digby, April 20, 1657.

† " In serie numerorum 1, 6, 30, 140, 630, etc. quæritur terminus medius ipsis 1 et 6 interponendus. Fieri autem hos terminos indicabam, ex continua multiplicatione numerorum $1 \times (4+2/1) \times (4+2/2) \times (4+2/3) \times$, vel etiam $1 \times 6/1 \times 10/2 \times 14/3$, quorum tam numeratores quam denominatores sunt Arithmetice proportionales." Dedication, *Arith. Inf.,* and Letter to Oughtred, Feb. 28, 1654/5 (Rigaud i, 85).

CHAPTER V

But, complained Fermat, how can one insert a mean between 1 and 6, when this series does not start at 1 ? " Il faut aussi remarquer ", he said, " que le raport des nombres de ladite progression n'arrive pas jusques au premier terme 1. ou plûtôt ne commence pas dés le premier terme, mais au second seulement, qui est sa borne . . . . Puis donc qu'il ne faut pas avoir égard au premier terme 1. qui n'a rien de commun avec les nombres de ladite progression ; mais aux autres seulement, et qu'ils augmentent à mesure, qu'ils approchent du premier terme 1. il s'ensuit, que le nombre, qu'on prendroit entre 1. et 4. 2/1 ou $\frac{6}{1}$ seroit plus grand, que ledit $\frac{6}{1}$ ou 6 " *. Thus if we consider the terms of the series subsequent to the first, which as Wallis points out are obtained by the continued multiplication of

$$4+\frac{2}{1},\ 4+\frac{2}{2},\ 4+\frac{2}{3},\ \text{etc.,}$$

any attempt to "interpolate" a term between the first and the second from the law of the series would lead to a result greater than the second term, which is manifestly absurd. This, maintained Fermat, was a serious defect, and Fermat was unquestionably right.

Again, continued Fermat, in his first proposition, Wallis had proposed a series of quantities commencing at 0, which proceeded in arithmetical progression, and he had sought the ratio of their sum to the sum of a like number of terms each equal to the last. The method which he had adopted to find this ratio was to consider the sums of different numbers of terms, in each case to find the required ratio, and hence infer a general rule. Against this method of procedure Fermat protested vigorously. " On pourroit proposer telle chose et prendre telle regle pour la trouver, qu'elle seroit bonne à plusieurs particuliers, et neantmoins seroit fausse en effect, et non universelle " †—*i. e.*, a rule which is valid in a number of cases is not necessarily so universally. Thus, argued Fermat, the fundamentals being unproved, how could the structure erected thereon be sound? Not only that, persisted Fermat, but this style of demonstration, which is founded upon induction rather than upon reasoning according to the principles of Archimedes, makes much

---

* " It must be noted that the ratio of the numbers of the progression in question does not reach the first term 1, or rather does not begin with the first term, but only with the second, which is its limit . . . . Since then no attention need be paid to the first term which has nothing in common with the other members of the said progression but only to the others, and as they increase as they approach the first term 1, it follows that the number taken between 1 and (4+2/1) or 6 would be greater than the given 6." Fermat, *Opera Varia*, p. 194.

† Fermat, *op. cit.*, p. 195.

difficulty for the novice. Finally, he urged that Wallis should " lay aside the Notes (for some time at least), Symbols or Analytick Species (now since Vieta's time in frequent use) in the construction and demonstration of Geometrick problems, and perform them in such method as Euclide and Apollonius were wont to do, that the neatness and elegance of Construction and demonstration, by them so much affected, do not by degrees grow into disuse " *.

To this Wallis replied that it would not have been a difficult matter to have evinced the truth of his discoveries by such demonstrations and deductions *ad impossibile* as had been made use of by the ancient geometers. But that would be to lose sight of the main object of the treatise, which, unlike those of the ancients, sought to demonstrate a *method of investigation*, not the discovery of new truths. Had not modern geometers more than once complained of the practice of the ancients, who appeared to be more studious to demonstrate a truth than disclose how they came by it ; was not he therefore deserving of praise rather than censure for his ingenuity ? †. Moreover, as to Fermat's assertion that the Method of Induction employed in the pages of the *Arithmetica Infinitorum* was not to be commended, and that he himself could give a demonstration " in lesser room ", Wallis stoutly challenged this assertion, claiming that his method was " a very good method of *Investigation* ; as that which doth very often lead us to the easy discovery of a General Rule ; or is at least a good preparative to such an one " ‡, of which statement there could be no better illustration than the early propositions of the *Arithmetica Infinitorum*. Wallis claimed that his induction was conclusive, " and the same may be said of all the Inductions which I make use of ; Which I always pursue so far . . . . till it lead me into a regular orderly Process ; and, for the most part (if not always) to an Arithmetical Progression ; in which I acquiesce as a sufficient evidence, when there is no colour of pretence, why it should not be thought to proceed onward in like manner. And without this, we must be content to rest at particulars (in all such kind of Process,) without proceeding to the Generalls " §. This last sentence indicates a most valuable feature of Wallis's

* " Sepositis tantisper speciebus Analyseos, Problemata Geometrica viâ Euclideanâ et Apollonianâ exequi, ne pereat paulatim elegantia et construendi et demonstrandi, cui præcipue operam dederunt veteres." *Algebra*, p. 305, and *Com. Epist.*, 46 (not 47 as Wallis says).

† " Grates siquidem ego potius expectassem, quam ut eo nomine criminis insimularer, quod aperte et sine fuco non modo quo perveneram sed et quibus passibus indicaverim ; nec (quod de aliis non pauci conqueruntur) pontem illum ipse demolitum iverim, quo ego flumen transieram." *Com. Epist.*, xvi, Wallis to Digby, Nov. 21, 1657.

‡ *Algebra*, p. 306.

§ *Ibid.*, p. 308.

mathematical work, namely, his cultivation of general methods. " Sure I am ", he wrote again in his *Algebra*, "that since the Introduction of such Methods, Mathematicks have been more improved in this present age, than they had been in many ages before. And such things as were (singly) wont to be looked upon as profound discoveries, are now (in Multitudes innumerable) by general Methods (of which I take this Arithmetick of Infinities to be none of the most contemptible) easily discoverable by a direct Calculation " *.

It is a curious fact that in an epoch such as this, characterized as it was by outstanding advances in mathematics, general methods found so little favour. Even such able mathematicians as Fermat did not always aim at general solutions, and of that vast body of mathematicians, even down to the time of Euler, who made mathematics such a potent instrument of physical research, scarcely any save Wallis encouraged the development of generalized methods. Not only do we find innumerable examples of this in the *Arithmetica Infinitorum*, they appear again and again in all his mathematical work, and not the least in his investigations into the Theory of Numbers. But despite the great service which he thus rendered to mathematics, many years were to elapse before the methods which he advocated were taken up and vigorously pursued.

Nevertheless, in spite of his spirited defence in the face of Fermat's objections, there can be no doubt as to the soundness of the latter's criticism of the Method of Induction employed by Wallis. Unerringly Fermat had put his finger upon a definite blemish in Wallis's work. The logic of the Method of Induction as used by Wallis was not flawless. Like many other mathematicians of the period he had fallen into the error of generalizing from a number of isolated cases. This procedure was not rigorous, and the fact that in his hands it frequently led to correct results could not justify it. "One will not deny", observes Cantor, "that to stop at a mere Induction, without deriving a general formula which serves the same as a support, that a bland *patet* (it is evident) does not satisfy the demands of modern mathematical rigour. One can just as little deny that Wallis first had the idea of proceeding to undertake the investigation after, one may even confidently say because, he foresaw the results to be expected " †. Such

* *Algebra*, pp. 297/8.

† " Man wird nicht verkennen, dass ein Stehenbleiben bei blosser Induction ohne Ableitung einer allgemeinen Formel, welche derselben als Stütze dient, dass ein kühl ausgesprochenes *patet*, es ist offenbar, nicht mit den heutigen Anforderungen mathematischer Strenge in Einklang zu bringen sind. Man wird ebensowenig verkennen, dass Wallis zur Anstellung seiner Versuche überhaupt erst überging, nachdem, man darf sogar getrost sagen, weil er die zu erwartenden Ergebnisse voraussah." *Cantor*, ii, p. 900/1.

criticism is severe, but it is nevertheless just. On the other hand, it does not give Wallis credit for the amazing patience which was a part of all his investigations.

Before long Wallis was involved in one of those unfortunate disputes concerning priority which were such a feature of the latter half of the seventeenth century. About the year 1661 Sir Kenelm Digby sent Wallis a rectification of a curve which had been accomplished by Fermat, who claimed it as a new discovery. Wallis promptly showed this curve to be identical in all but name with that rectified by William Neile in the July or August of 1657, an account of which Wallis, in his *De Cycloide* (1659), page 91, as well as many other notable writers, had already published. Moreover, as Neile readily admitted, the foundation of the method was contained in the pages of the *Arithmetica Infinitorum*, a work which Fermat had undoubtedly seen.

Now Fermat's dissertation in which his discovery was first published appeared in 1660 with the title *De Linearum Curvarum cum Lineis Rectis Comparatione*, and the curve whose rectification he had effected was, to use his own words, " una ex infinitis parabolis . . . . in qua Cubi applicatarum ad Axem, sunt inter se, ut quadrata portionum Axis ", *i. e.*, the curve, the cubes of whose ordinates are proportional to the squares of the abscissæ. Moreover, averred Wallis, Fermat had declared at the end of his fifth proposition : " ex una hac curva derivari et formari alias numero infinitas non solum ab ipsa sed inter se specie differentes quæ tamen singulæ rectis datis æquales esse demonstrentur ", *i. e.*, from this one curve are derived and formed others infinite in number, differing not only from this, but also from each other in type, which are, however, separately shown to be equal to a given right line *.

" But ", complained Wallis, " this curve is no other than that of Mr. Neil, which is, as I have shown before, a parabola whose ordinates are in the subtriplicate of the duplicate proportion of the intercepted axes. And those innumerable, which he says are different curves from that are indeed but the same curve (or parts of the same curve) beginning at different parts thereof". But a reply more typical of Wallis is to be found in the *Algebra*, where we find the following illuminating passage :—

" I will not disparage *Mons. Fermat's* Invention herein ", wrote Wallis, " nor his Demonstrations thereof. But allow the

* What Fermat actually wrote was : " Quid enim mirabilius quam ex una hac curva derivari . . . . esse demonstrentur ", *i. e.*, " For what is more wonderful than that from this one curve are, etc.". The passage occurs at the end of the *fourth* proposition, and not the fifth as Wallis asserts. Fermat, *Opera Varia*, p. 97.

Invention to be very Ingenious, and his Demonstrations to be good and full. (Save that he takes those to be so many different sorts of Curves, which are indeed all the same.) Nor will I impute it as a fault in him, that others had done the same thing before him : Or that he had (or might have had) the first hints of it from my *Arithmetick of Infinities*, (which I am sure he had read) " *.

Wallis made no mistake here. Neile's rectification was known at least three years before Fermat's appeared, and even if we allow that mathematicians were still working in comparative isolation, and that their results were but tardily disseminated, it is unlikely that a scholar of Fermat's sagacity would have long remained in ignorance of an achievement of such distinction. " The thing was so notorious, and known to so many (being then made publick to that Society at *Gresham* College) ", declared Wallis †. Nine years later Wallis wrote to Collins : " The whole mystery of rectifying curve lines is laid down in my Scholia ad Prop. 38, *Arithmetica Infinitorum*, and further prosecuted in my Cycloide, page 90. And I have not yet seen anything of that kind but what are but particular instances of what is there delivered in general " ‡. In his *Algebra* he wrote : " And I do not at all doubt but this notion there hinted, gave the occasion (not to Mr. Neil only, but) to all those others (mediately or immediately,) who have since attempted such Rectification of Curves, (nothing in that way having been attempted before) " §. Though these can be dismissed as further examples of Wallis's passion for exaggeration, it was not beyond a scholar of Fermat's extraordinary intuitive faculty to penetrate behind the hint which Wallis had given in the Scholia mentioned.

France, however, was not the only nation which tried to deprive the Englishman Neile of his claim to priority in the matter of curve rectification. Huygens, in his *Horologium Oscillatorum* (Paris, 1673), ascribed this invention to John Heuraet of Harleem in 1657 ‖ Wallis sharply refuted this in a letter to Oldenburg, Oct. 4, 1673, maintaining that besides Neile's rectification of 1657 many other Englishmen, notably Lord Brouncker and Sir Christopher Wren, had effected similar rectifications, and that one of these, namely Wren's rectification of the cycloidal curve in 1658, was known throughout France and Holland. This is abundantly confirmed

* *Algebra*, p. 297.
† *Ibid.*, p. 293.
‡ Wallis to Collins, Oxford, Sept. 8, 1668 (Rigaud, ii, p. 496).
§ *Algebra*, p. 298.
‖ *Horologium Oscillatorum*, Prop. 9, part iii, pp. 71, 72. Huygens, *Œuvres*, vol. xviii. La Haye, 1934.

by Wren and Brouncker in letters to Secretary Oldenburg. These letters were afterwards published in the *Philosophical Transactions*, and they emphasize, as nothing else can, the need of official scientific journals—a need which still for many years to come had to be met by private correspondence. Indeed, not the least important aspect of the many societies which were springing up over Europe was the fact that their respective publications constituted an international cockpit for diverse scientific opinion. Most of the unhappy quarrels as to priority, and the mutual charges of plagiarism amongst scholars, arose through genuine misunderstandings which would have been avoided had there been any recognized means of publication.

The correspondence seems to have had the desired effect, inasmuch as Huygens acknowledged his error, " and did expressly give us leave to print what we should think fit for reasserting it to Mr. Neil " *. According to Wallis the true history of the early attempts at rectification were :

(1) 1657 †. July or August. Neile rectified the semi-cubical parabola by a method based upon the principles outlined in the *Arithmetica Infinitorum*. This was subsequently published by Wallis in his *De Cycloide*, p. 90 (1659).

(2) 1658. Wren showed the curve of the cycloid to be quadruple its axis.

(3) Heuraet hit upon the rectification of the same curve, an account of his achievement being published by Francis van Schooten in the first part of his *Geometria Cartesiana* ‡.

(4) 1660. Fermat gave an account of the rectification of the same curve that Neile had rectified three years earlier, subsequently publishing an account in his treatise *De Linearum Curvarum cum Lineis Rectis Comparatione*.

* *Algebra*, p. 293.

† In a letter to Collins, Oxford, Sept. 27, 1673 (Rigaud, ii, 586) Wallis gives this date as 1653. " Mr. Neale gave a straight line equal to a crooked one which was done about June or July 1653." This is a blunder unusual in Wallis.

‡ Montucla (ii, art. viii) is strongly of the opinion that Heuraet could not have seen the work of the Englishman.

CHAPTER VI.

APPOINTMENT AS *CUSTOS ARCHIVORUM*.
THE QUARREL WITH DOCTOR HOLDER.

MEANWHILE Wallis's talents had earned for him still further recognition, for on February 17, 1657/8, he succeeded Gerard Langbaine as *Custos Archivorum* to the University. The appointment, however, does not appear to have met with very general approval. Of Wallis's contemporaries there were not a few who were only too eager to attribute his rapid advancement to a decided inclination to trim his sails to the prevailing political winds, and the views expressed by Aubrey in his *Lives of Eminent Men*, which was eventually published in 1813, were typical of many. " In 1657 ", declared that well-known antiquary and universal gossip, " he gott himselfe to be chosen (by unjust means) to be *Custos Archivorum* of the University of Oxon, at which time Dr. Zouch had the majority of voices, but because Dr. Zouch was a malignant, (as Dr. Wallis openly protested, and that he had talked against Oliver) he was putt aside. Now, for the Savilian Professor to hold another place besides, is so downright against Sr. Hen. Savile's Statutes that nothing can be imagined more, and if he does, he is downright perjured. Yet the Dr. is allowed to keep the other place still " *. Hardly less virulent is the account given by Anthony à Wood in the *Athenæ Oxonienses* : " The famous Dr. *Rich. Zouch*, who had been an Assessor in the Chancellours Court for thirty years or more, and was well vers'd in the Statutes, Liberties, and Privileges of the University, did, upon great intreaties stand for the said place of Antiquary or *Custos Archivorum* thereof, but he being esteemed a Royalist, Dr. J. W. was put up and stood against him, tho altogether uncapable of that place, because he was one of the Savilian Professors, a *Cambridge* man, and a stranger to the usages of the University. At length by some corruption, or at least connivance of the Vice Chancellour, and perjury of the Senior Proctor (Byfield), W. was pronounced elected " †. Henry Stubbe, who had already cultivated an intense dislike for Wallis, whom he had accused of having deciphered the King's Cabinet taken at Naseby " to the ruin of many loyal persons " ‡, and who had vigorously supported

---

* Aubrey, ii, p. 569.
† Wood, ii, col. 415.
‡ *Ib.*, ii, col. 414, 415.

Hobbes in his quarrel with Wallis, now rushed into the arena with a tract: *The Savilian Professour's Case stated. Together with the several Reasons urged against his Capacity of standing for the Publique Office of Antiquary in the University of Oxford; Which are enlarged and vindicated against the Exceptions of Dr. J. Wallis* (Lond., 1658). In spite of these recriminations, however, Wallis was duly elected, and, if we are to rely upon his own evidence, he carried out his duties conscientiously enough. " I think it will be acknowledged ", he wrote, " that I have done the University considerable services; I am sure it hath been my endeavour to do so, when I have been employed by them " *. Much more convincing testimony, however, is to be found in Doctor Gregory's *Account of Wallis's Life*, where we read the following : " In 1657 he was chosen Custos Archivorum. In this last place he put the records and other papers belonging to the University that were under his care into such exact order, and managed its lawsuits with such dexterity and success that he quickly convinced all, even those who made the greatest noise against this election, how fitt he was for that post. As for the other [Savilian Professorship] he made so great a figure in it that he justly gott the name of one of the greatest geometers of that age, so famous for the improvement of that Science " †.

His dual office, however, could not restrain his passion for controversy, and within a few weeks of this last appointment he was drawn into the vortex once more. The occasion of this was the prize questions on the cycloid proposed by Pascal (July 17, 1658), to which reference is made elsewhere. But not even this, nor the disputes with Fermat and with Hobbes, which were still raging fiercely, appears to have checked the output from his able pen, for in 1659 he published his treatise *De Cissoide et Corporibus inde Genitis*, to which he prefixed another: *De Cycloide et De Corporibus inde Genitis*. In this latter he gave a demonstration of Neile's rectification of the semi-cubical parabola, and he took the opportunity of again calling attention to the extent to which his own work had prepared the ground for Neile's achievement. These two treatises afterwards appeared in his celebrated work *De Motu*.

The year 1660 saw the Restoration of the exiled Stuart Charles II, and Wallis, in spite of his adherence to the Cromwellian party, was received very graciously by the royal monarch, to whom the pose of universal patron of learning seems to have been particularly attractive. Not only were his two appointments confirmed, he was actually admitted one of the King's Chaplains in Ordinary. In 1661 he was appointed one of the Divines chosen to revise the

* Hearne, p. 169.
† Gregory, *Account of Wallis's Life* (Smith MS. 31). Bodleian.

Common Prayer Book, and the following year the degree of D.D. was conferred upon him. Later he complied with the terms of the Act of Uniformity, and he continued a zealous Conformist to the Church of England until his death.

In the January of 1661/2 Wallis's extraordinary versatility manifested itself in yet another direction, for in that year he undertook to teach to speak distinctly one who from childhood had been deaf and dumb. This was Daniel Whaley, and so remarkable was the success which he attained that, according to Wallis, Whaley, on May 21, 1662, " did in the presence of the Society (to the great satisfaction of the Company) pronounce very distinctly enough such words as were by the Company proposed to him, and though not altogether in the usual Tone or Accent, yet so as easily to be understood " *. Shortly afterwards Wallis was called upon to perform a like service for one Alexander Popham, a youth who had also been deaf and dumb from the cradle. But even in the performance of such an act of humanity he could not avoid the fetters of controversy, and soon afterwards he found himself involved in another long and bitter dispute. For Doctor William Holder, of Cambridge, " a great virtuoso and a person of many accomplishments ", a man with whom Wallis had more than once crossed swords, had been responsible for the early training of the deaf boy, and it seems unreasonable to doubt that the latter had derived substantial benefit from Holder's training. But when, in 1670, Wallis was invited by the Royal Society to lay before them an account of the cure, no mention whatever was made of Holder. Even when the account was published in the *Transactions* for that year the omission was not repaired. For eight long years Holder nursed a genuine grievance ; finally his pent-up anger gave itself expression in *A Supplement to the Philosophical Transactions of July* 1670, *With some Reflexions on Dr. John Wallis, his Letter there Inserted* " (Lond., 1678). In this Holder accused Wallis of assuming the glory of having taught young Popham to speak. As will be seen from the following extract Holder's paper was written with little restraint, but then it can hardly be denied there was considerable provocation for his charges against Wallis.

" The Author ", he wrote, " is not so much concerned to be righted to his Title against an Invader, as to express his just Resentment of the subtle practices, which have been contrived to abuse and mislead the Reader with False shews, somewhat resembling Truth . . . . He [Wallis] had a long aking-tooth, to joyn to his other Trophies, that which he saw performed by Doctor *Holder*, and silently passed over, viz. The finding a successful way of teaching Dumb and Deaf persons to speak ". Holder then

* Wallis to Oldenburg, July 11, 1670. (*Royal Society Guard Books.*)

went on to refer to the Doctor's skill in thus penning and spreading his own fame, " in large Characters ingraven by himself. For that purpose the Transactions were his common Market, and a new Book upon the Anvil, if he could find way and leave to croud himself in, was a Fair for this Merchant of Glory. And when he got a hint, (for which he always lay in wait) of any considerable new Invention or Improvement, presently comes out an Epistle or small Tractate of Doctor *John Wallis* upon that subject, to entitle it to himself " *. Wallis did not hesitate to reply, and in a Tract entitled *A Defence of the Royal Society, and the Philosophical Transactions, particularly those of July* 1670, *in Answer to the Cavils of Doctor William Holder* (Lond., 1678) he treated Holder with unusual severity. Oldenburg, in whose pages the dispute raged, realized the justice of Holder's indignation. "The publisher", he wrote in the Preface to the above *Supplement,* "to avoid partiality on his part, (though but in appearance) hath thought fit to publish this ensuing Narrative of what hath been done in this kind by Doctor Holder, as it is handed to him by the Author himself". He then went on to say : " Dr. *Holder* .... could not but wonder, and almost pity to see a person of so good Learning, endowed with so many excellent things to make him very considerably reputed in the world ; yet by too much Greediness of Fame, suffer himself to be tempted to the vanity of using such ways of begging and borrowing Reputation ". Wood, in whose mind irritation at Wallis's appointment as *Custos Archivorum* still persisted, reviewed the affair some years later, and bitterly accused Wallis of vainly taking the glory to himself without acknowledging what had been done before him. He referred to the *Supplement* with some asperity, adding : "This last was written by him [*i. e.,* Holder] to vindicate himself that he had taught Mr. *Popham* to speak, which Doctor *Wallis* in the said Letter did claim to himself : Whereupon soon after, Dr. *Wallis* (who, at any time can make black white, and white black, for his own ends, and hath a ready knack of sophistical evasion as the writer of these matters doth know full well) did soon after publish an answer " †. It is fairly obvious that both Wallis and Holder had contributed much to the boy's recovery. Holder's methods were described in a paper read to the Society on February 25th, 1668, entitled *Elements of Speech. An Essay of Inquiry into the Natural Production of Letters* " (Lond., 1669), by William Holder, D.D., F.R.S., a paper which gives evidence of profound research, and which was highly commended in the *Transactions* (p. 958) for that year, where it was described as " a well considered and useful Tract ". Wallis had discussed

† Holder, *Supplement to the Phil. Trans. of July* 1670 (London, 1678).
‡ Wood, *Fasti,* ii, col. 815, 816.

CHAPTER VI

the theory of this art in a treatise, *Grammatica Linguæ Anglicanæ, cui præsigitur De Loquela (sive De Sonorum Omnium Loquelarum Formatione)* (1653), in which were described in detail the various modes of production of articulate sounds. But Wallis's interest in the practice of this art can be traced to a much earlier date. In 1661 he wrote to Boyle : " I am now upon another work, as hard almost to make Mr. *Hobbes* understand a Demonstration. It is to teach a person deaf and dumb to speak " *, and if we are to rely on his own testimony he had achieved success in this art on other occasions. " I have since that time ", he wrote to Doctor Smith, " taught divers persons (some of them very considerable) to speak plainly and distinctly who before did hesitate " †. It is interesting to observe that quite likely both Holder and Wallis had profited by the pioneer work of Wilkins, who many years earlier had shown how the epiglottis, larynx, aspera arteria (*i. e.*, trachea) and the œsophagus help in the production of the various sounds in speech.

The next mathematical work which Wallis contributed was his *Cono-Cuneus*, or The Shipwright's Circular Wedge, which was written in 1662 at the request of Sir Robert Moray. This arose out of a discussion as to the nature of the various solids which were constructed according to the following plan :—

" On a plain Base, which was the Quadrant of a Circle, (like that of a Quadrantal Cone or Cylinder) stood an erect Solid, whose Altitude (being arbitrary) was there double to the Radius of that Quadrant ; and from every Point of its Perimeter streight Lines drawn to the Vertex, met there, not in a Point, (as is the Apex of a Cone) nor in a parallel Quadrant (as in a Quadrantal Cylinder) but in a streight Line or sharp Edge, like that of a Wedge, or Cuneus " ‡. Now it was argued that by taking different sections of such a solid much information useful in the building of ships might be gathered, The method hitherto adopted had been by actual section of the models. In this treatise Wallis showed how these sections could be demonstrated *in plano* without actual section, in much the same manner he had used in illustrating the properties of the Conic Sections.

Every fresh manifestation of his genius served to establish even more firmly Wallis's reputation as one of the most distinguished scholars of the day. He rapidly became an outstanding figure of the Royal Society, and few matters came up for the consideration of its members which he was not competent to discuss. This, mated to a resolute courage, rendered him a valuable member of

* Wallis to Boyle, December 30, 1661. (Boyle, v. p. 511.)
† Wallis to Doctor Smith, Jan. 29, 1696/7 (Smith MS. 31, Bodleian).
‡ *Cono-Cuneus*, p. 1.

the Society whose reputation he strove to advance, not merely by contributing papers of his own on a multiplicity of subjects, but also by giving at their request an account of such as were received by the Society. By this time the Royal Society had become a clearing-house for information regarding current research, and in no branch of Wallis's work did he exhibit greater thoroughness than in the task to which that learned body had appointed him. Yet such were the conditions of the age, in no branch of his work did he make more enemies. "I have yet been so carefull all along", he wrote to Oldenburg, " in what I say as to speak only my own opinion, without declaring for others who's opinions I know not, lest while I go about to right myself I should wrong them "*. Wallis did not always practise what he liked to preach. He was not cast in the same mould as Newton, whose inherent fear of controversy was in the last resort hardly less disastrous than Wallis's passion for it, and it is greatly to be deplored that one who played so great a part in the development of learning should not have shown himself capable of rising above the petty rancours which seem to have been such a characteristic of seventeenth century science.

For Wallis was of a highly contentious disposition. His correspondence unhappily leaves no room whatever for doubt on that point. No man ever scorned personal popularity more completely than he, and his unyieldingly logical temperament and his ineradicable hatred of compromise often made him appear to have very little desire to understand the point of view of those from whom he differed, and made him appear so relentless in passing judgment that sometimes he seemed to forget that he too could err. Embittered by his personal quarrel with Fermat he became more and more intemperate in his attacks upon the French, where more than once his criticism led to vigorous protests. His indictment of Du Laurens, for example, whose *Specimina Mathematica* appeared in 1667, is so virulent and unrestrained that it merely defeats its own purpose. For after showing it to be a treatise of appalling crudity, he added : "Dixeram partem magnam ex Oughtredi meisque scriptis desumptam videri " †—a double charge of incompetence and plagiarism which could convince no one. The consequences of this unfortunate characteristic were realized by Halley, who begged that "all dissatisfactions and uneasiness be at an end,

---

* Wallis to Oldenburg, Nov. 14, 1668. (*Royal Society Guard Books*.)

† "I had said a greater part seemed to be taken from my writings and from those of Oughtred." Wallis to Oldenburg, July 2, 1668. (*Roy. Soc. Guard Books*.)

CHAPTER VI

they being hardly consistent with a Philosophicall Genius " *. Indeed, so prone was Wallis to rush headlong into controversy that a dispassionate retrospect of his activities in this direction points to the fact that the Society's choice of Wallis for this important task was of questionable wisdom. Not that he was academically ill-equipped ; on the contrary, even in that illustrious age it would not have been easy to find one expert over such a wide range of subjects as was Wallis. But he lacked the patience, the tact, which were indispensable to one who was little less than the mouthpiece of the Royal Society. A note of pedantic rectitude was never long absent, from which even his fellow-countrymen were not immune. James Gregory, for example, whose *Vera Circuli et Hyperbolæ Quadratura in Propria Sua Proportionis* (1667) earned a laudatory review in the *Transactions*, and which Collins spoke of as " deserving much applause " † was denounced by Wallis for " betraying his ignorance and unskilfulness in the nature of a Demonstration " ‡. Such scathing attacks as we meet again and again in his letters come ill from the pen of one who occupied such a dominating position in the world of learning.

About the year 1666 Wallis turned his attention to the phenomenon of Tides. Many attempts had already been made to explain the apparent irregularities in the sea's ebb and flow, and many curious suggestions had been advanced. Galilei, it will be recalled, had attributed them to the different velocities of different parts of the earth. Wallis's letters, particularly those to Oldenburg (May 19, 1666) and to Boyle (April 25, 1666) show that he was devoting much time to an elucidation of the phenomenon, and the fruits of his investigation eventually appeared in a treatise, published in 1668, under the title *De Æstu Maris, Hypothesis Nova*. The essential parts of this had already been laid before the Royal Society on August 6th, 1666. Wallis's treatise was remarkable for the sagacious assumption that the earth and the moon might be regarded as a single body concentrated at the common Centre of Gravity. To account for the variation of the tides from Spring to Neap he assumed that this common Centre of Gravity was not stationary, but moved. Though his calculations were exact, he did not consider the action of the sun ; this was left for Newton, whose hypothesis eventually superseded that of Wallis.

Meanwhile Wallis had been gathering together the fragments of what had passed in those days for mechanics. The outcome

* Halley to Wallis, July 9, 1686. (MacPike, *Correspondence and Papers of Edmond Halley*, 1932, p. 67.)
† Collins to Gregory (undated). Rigaud, ii, p. 174.
‡ Wallis to Oldenburg, Nov. 27, 1668. (*Royal Society Guard Books*.)

of his patient researches in this field was the treatise *De Motu*, which was finally published in 1670. In some respects this monumental work may be adjudged his masterpiece; certainly it gives evidence of profound investigation into a somewhat neglected subject. This becomes plain when we examine the condition in which this subject was to be found when Wallis first directed his energetic mind towards it, and when it is noticed, as it cannot fail to be, how near the Integral Calculus he actually approached in its pages.

CHAPTER VII.

The *Mechanica, sive Tractatus de Motu*.

It is an unmistakable sign of Wallis's greatness that, unlike most of his predecessors and his contemporaries, his mathematical investigations were not restricted to the field of pure mathematics. The related subject mechanics was no less vigorously prosecuted by him ; indeed, it is scarcely too much to say that of all the contributions which this versatile genius made to the progress of science few perhaps ought to rank higher than his vast and unwearied researches in mechanics. Yet his greatness here is not to be measured by the extent of the ground he opened up. This was but slight. What rendered his labours of such inestimable value was that he gathered together the mass of crude and ill-digested notions which in those days passed for mechanics, and out of the medley evolved something like a coherent system. This was no mean achievement. For it cannot be too strongly urged that mechanics had never shared in that general awakening which was such a distinctive feature of the history of science during the later Middle Ages. Whilst almost every other branch of learning was being pursued with conspicuous zeal, few scholars seem to have considered mechanics worthy of their notice. Mathematics, for example, had been cultivated almost without any breach of continuity right from antiquity ; the mathematician of the seventeenth century could at least be sure of solid foundations upon which to build. With mechanics, however, vastly different conditions prevailed. Even after Archimedes, in his *Equiponderance of Planes*, had indicated the right path, the science remained quiescent for nearly 2,000 years. During the Middle Ages the elucidation of the fundamental problems of statics and dynamics attracted the notice of singularly few men of science, and even these few seemed to have been quite content with nothing more erudite than prolix commentaries upon the writings of Aristotle, and to accept the vaguest speculations so long as they accorded with the canons of the great Stagyrite philosopher. It is true that during the fifteenth and sixteenth centuries we meet a few distinguished mathematicians, as, for example, Nicholas of Cusa, Cardan, Tartaglia, who devoted some attention to mechanics, but their writings reveal little originality, and their speculations, unsupported as they were by experimental verification, were of little permanent value *, even when they happened to light upon the truth, and that was not frequent. Cardan, for example, in his treatise *Opus Novum De Proportionibus Numerorum, Motuum, Ponderum* (Basle, 1570), investigated the relation between the force necessary to hold a body on an Inclined Plane and its weight, and he made the force proportional to the angle of the plane. This conclusion was based upon the knowledge that when the angle is 0 the force is zero, and when the angle is 90° the force is equal to the weight of the body. But it is a striking commentary upon the casual way in which these questions were examined that a scholar of Cardan's ingenuity could not have seen further than this. It is therefore no exaggeration to assert that the dawn of the seventeenth century saw the fundamental ideas of motion, and even the elementary principles of statics, still in a chaotic state; and if further testimony is needed it is to be found in the almost incredible fact that right down to the threshold of the century the crude Aristotelian notions regarding the descent of heavy bodies persisted. Even when these were disproved there were still many who clung to the belief that the distances described by a falling body were proportional to the natural numbers 1, 2, 3, 4, etc. The motion of projectiles had after a fashion been investigated, and some curious speculations had gained currency. Motion, according to sixteenth century doctrines, was either *natural*, as, for example, the circular motion of the stars and the descent of heavy bodies, or *violent*, i. e., so contrary to nature that it could not subsist without the continued application of a *force motrix*. The confusion of thought which prevented mathematicians from perceiving the difference between producing motion and preserving it was fatal to all attempts at an elucidation of the most elementary problems. Why did a stone remain in flight for a long time after it had been released? It was, asserted the docile followers of Aristotle, because the air which followed from behind continued to give it motion; consequently the original direction persisted until at length the body fell vertically. The idea that a body could have two motions simultaneously was quite foreign to the seventeenth century mind. Tartaglia realized that the path of a projectile was a curve of some sort, and he compounded it out of a straight line, the arc of a circle, and the vertical tangent to the arc, and

* An exception, perhaps, might be made in the case of Leonardo da Vinci, whose works, had they been published, would have advanced the science to the position it occupied a century later. But as his works remained unpublished until the close of the eighteenth century, they can hardly have affected the progress of the science, even granting, as Duhem has suggested, that Cardan helped himself to them long before they were made public.

CHAPTER VII

although he actually asserted that a projection at an angle of 45° gave the maximum range, he could hardly have lit upon this truth but by hazard.

The doctrine that " every body seeks its natural place " found popular acceptance, and was current even as late as 1634. " Pesanteur est la force ", wrote Albert Girard, the Editor of the Elzevir edition of Stevin (Leyden, 1634), " qu'une matiere demonstre à son obstacle, pour retourner en son lieu " *. That such doctrines should have been accepted speaks eloquently of the deplorable condition of mechanics in even that energetic age. Scholars were still a long way from realizing that, contrary to what Aristotle and his all too servile followers had maintained, a body in motion would continue at the same speed and without change of direction unless impeded. Speculations such as characterized sixteenth century mechanics were thus far too vague to serve as a starting-point for real progress, and until the perception of the true grounds of the philosophy of mechanics had been realized progress was impossible. This was Wallis's great achievement. By co-ordinating and extending the little that was known, by his constant appeal to experiment, and by his rigid application of the laws of mathematics to mechanics, he gave an impulse to the development of the subject which it had not known for centuries.

The earliest attempts at clearing up some of the confusion, however, appear in the work of Guido Ubaldo del Monte (1546–1607), whose *Mechanicorum Liber* (Pesaro, 1577) contained some interesting speculations on the principle of the Lever. He held the correct view concerning the ratio of the forces applied to the arms of a bent lever to keep it in equilibrium, and he used this to establish the true theory of the Balance. He was not so fortunate, however, in his investigations on the Inclined Plane and the Screw, which, according to Cantor †, he failed to understand. Perhaps more enduring were the researches of Benedetti (1530–1590), whose chief services to the young and feeble science were his critical scrutiny and correction of the doctrines of Aristotle. In his *Diversarum Speculationum Mathematicarum et Physicarum Liber* (Turin, 1585) he investigated several mechanical problems. He attributed centrifugal force to the tendency of a body to move in a straight line, and that in itself was a considerable improvement on the notions then current, and showed a clear conception with regard to motion which even Galilei was long in acquiring. He realized that falling bodies were accelerated, and he explained the acceleration as due to the summation of the impulses of gravity,

---

* Stevinus, *Œuvres*, vol. ii, p. 434.
† Cantor, ii, p. 568.

though he fell into the curious error of assuming that the velocity of descent was proportional to the density. He avoided the errors of his predecessors in explaining the motion of projectiles, which he vaguely maintained, was due to a *virtus impressa* ; but he never attained perfect clearness, and he fell a victim to the illusion that the action of gravity was annulled by the velocity of projection.

In the writings of Benedetti, however, appears the notion still vague and ill-defined, but none the less valuable, namely, that to make any progress mechanics must be grounded upon mathematics. " It is the beginning of a mechanics founded on Geometry ", observes Cantor, " which is revealed in these considerations. Mechanics gradually ceases to collect mere empirical results or what was worse, to noise abroad philosophical deductions, unconcerned whether they agree with experience or contradict it. Mechanics begins to be a chapter of mathematics " *.

A work of much greater erudition was the *Hypomnemata Mathematica* (Leyden, 1608) of Stevinus of Bruges (1548–1620), and it is probably no exaggeration to say that during the eighteen or nineteen centuries that separated Archimedes and Stevinus nothing of solid importance had been contributed to mechanical theory. In this treatise Stevinus corrected many of the errors then current ; more than that, his investigations enriched the science of statics with a large number of discoveries, chief of which was the principle of the Triangle of Forces. His work, too, contains the germ of the important principle of virtual displacements, foreshadowed by Aristotle and afterwards used so effectively by Wallis. He showed that in a system of pulleys in equilibrium the products of the weights into the displacements they sustain are respectively equal ("Comme l'espace de l'agent, à l'espace du patient ; Ainsi la puissance du patient, à la puissance de l'agent ") †. He elaborated the already known rule that two forces P and Q supporting a weight W (fig. 11) are inversely proportional to their distances from the weight, and although traces of an investigation of this sort are to be found in earlier writers, notably Varro (*Tractatus de Motu*, Geneva, 1584), the reasoning of Stevinus was vastly superior to anything that had preceded it.

Stevinus seems to have been the first to recognize the correct

* " Es ist der Anfang einer geometrisch begründeten Mechanik, der sich in diesen Betrachtungen enthüllt. Die Mechanik hört allmälig auf, blosse Erfahrungssätze zu sammeln, oder, was noch schlimmer war, philosophisch abgeleitete Behauptungen in die Welt zu schleudern, unbekümmert darum, ob sie zur Erfahrung passen oder ihr widersprechen. *Die Mechanik beginnt ein Kapitel der Mathematik zu werden.*" Cantor, ii, p. 570.

† Stevinus, *Œuvres*, Deuxième Partie, De la Statique, p. 509.

relation between the weight on an inclined plane and the force necessary to support it. Moreover, his hydrostatics was no less original. He examined the pressure on the surfaces of vessels which contained the liquids, and in no small measure he prepared the ground for Torricelli.

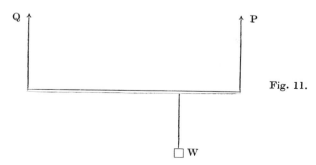

Fig. 11.

The work of Stevinus was taken up by Galilei, whose attempts to substitute a system of mechanics based upon experimental observation for the crude speculations of the Scholastics constituted his outstanding achievement. In his hands the science of mechanics began to assume a new aspect. His most enduring work belongs to the province of dynamics; indeed, it is hardly too much to say that his *Discorsi e Dimostrazioni Matematiche intorno a due nuove Scienze* (Leyden, 1638) initiated a complete reform in that neglected branch of mechanics. Scarcely less important than the experiments by which he overthrew the Aristotelian doctrine of falling bodies were his investigations upon the motion of projectiles and his attempts, unsuccessful though they were, to unravel the laws governing the impact of bodies. In Galilei's writings also appears the statement that a body in motion will continue in a straight line with undiminished speed, and that a body under the influence of two oblique impulses will follow the diagonal of the parallelogram whose sides are as these impulses. Both these form the subjects of some penetrating observations by Wallis less than half a century later.

The reforms which Galilei introduced were taken up and developed by two of his pupils—Castelli (1577–1644), whose *Della Misura dell' acque correnti* (Rome, 1628) created a new department of hydraulics, and Torricelli (1608–1647), whose *Opera Geometrica* (Firenze, 1644) was remarkable for some noteworthy contributions to the science of hydrostatics. In the section *De Motu Gravium naturaliter Descendentium et Projectorum, Libri duo* Torricelli added several new and important propositions to those which had been given by his master on projectiles. He seems also to have been the first to establish the principle applicable to all statical problems, that if two weights are so connected that when placed together in any position their common Centre of Gravity neither ascends nor descends they are in equilibrium in all these positions.

Meanwhile France, already Italy's rival in mathematical discoveries, seems to have become no less active in opening up new avenues in mechanics, and, stimulated no doubt by the achievements of Galilei, certain French mathematicians began at last to turn their attention to mechanical problems, and to cultivate the subject no less assiduously. *L'Harmonie Universelle* (1637) of Mersenne (1588–1648) and the *Essais Mécaniques* * of Roberval (1602–1675) contained demonstrations upon important branches of mechanics and hydrostatics, such as the vibration of rods, the resistance of solids, and the flow of water. Roberval used extensively the principle which to-day is known as the Principle of Moments, and although he displayed sound judgment in many of his investigations he was unable to reach perfect clearness with regard to the idea of *Force*. Mersenne's contributions are also contained in his *Cogitata Physico-Mathematica* (Paris, 1644), an unequal work which Montucla describes as " un océan d'observations de toute espèce, parmi lesquelles il y en a un grand nombre d'assez puériles " †.

In the writings of Descartes (1596–1650), particularly his *Lettres*, the indistinctness of fundamental notions is still apparent. An excessive confidence in his metaphysical speculations, combined with a reluctance to submit his investigations to experimental verification led him into some curious errors, and consequently little of his mechanical work was of permanent value. " We find in fact in his rules ", says his fellow countryman Montucla, " every kind of mistake, conjectural principles, contradictions, ... a tissue of errors which would not be worth discussing were it not for the fame of their author " ‡. His well-known Theory of Vortices, for example, succumbed before the criticism of Newton, who showed it to be impossible as a dynamical system.

Descartes' physical theory rests upon the metaphysical conclusions of his *Meditations*. According to him the primary physical conception is *Motion*, and this, having been created by God in the beginning, is unalterable in quantity. (" Deum esse primariam motus causam ; et eandem semper motus quantitatem in universo

* *Anc. Mém.*, Paris, T. vi.

† Montucla, ii, p. 207.

‡ " Nous trouvons effectivement dans ces règles toute sorte de défautes, principes hasardés, contradictions .... un tissu d'erreurs que ne mériteroient pas d'être discutées sans la célébrité de leur auteur." Montucla, ii, pp. 209/210.

conservare ")*. He enlarged upon this, declaring that "though motion is only a condition of moving matter, yet there exists in matter a definite quantity of it, which in the world at large never increases or diminishes, although in single portions it changes ". Thus the only circumstance which Science has to consider is the transference of Motion from one particle to another, and the change of its direction ; we may change the direction but not the quantity of Motion. Force is nothing more than one of those obscure conceptions which originate in some remote fashion in the sense of muscular power ; consequently, the whole idea may well vanish from the earth.

Extended body has no limits. The Universe is full of matter ; there can be no such thing as vacuum.

Descartes did not recognize that a body possessed *Inertia*. " I do not consider that there is any Inertia or inherent inactivity in bodies ", he wrote, " any more than does M. Mydorge. Further I believe that when a man walks, he causes the whole mass of the earth to move by however little it may be, but at the same time I agree with M. de Beaune that the greatest bodies being impelled by an equal force, as the greatest ships are by the same wind, always move more slowly than the others, which should perhaps be sufficient grounds for his reasons without calling in a natural Inertia which cannot be proved " †.

Descartes' speculations led to ten laws of nature, of which the first two are not wholly dissimilar from those usually associated with the name of Newton, whilst the rest, which deal mainly with Impact, are inaccurate. Nevertheless, Descartes was one of the first to realize that in Impact the communication of Motion was governed by laws, and his attempts to unravel these were certainly ambitious. He restricted his enquiry, however, to bodies absolutely hard, and, following the principles enunciated above, he deduced the following " laws " :—

(1) If two equal bodies strike one another with the same speed they will be reflected without loss of speed.

* *Principia Philosophiæ*, Pars Secunda, Elzevir, 1656, p. 37.

† " Je ne reconnois aucune Inertie, ou tardiveté naturelle, dans les cors, non plus que M^r. Mydorge. Et croy, que lors seulement qu'un homme se promene, il fait tant soit peu mouvoir toute la masse de la terre, ... Mais je ne laisse pas d'accorder à M^r. de Beaune, que les plus grands cors, estant poussez par une mesme force, comme les plus grands Bateaux par un mesme vent, se meuvent tousjours plus lentement que les autres ; ce qui seroit peut-estre assez pour établir ses raisons, sans avoir recours à cette Inertie naturelle, qui ne peut aucunement estre prouvée." *Lettres de M^r. Descartes au R. P. Mersenne.* Lettre xciv (ii, pp. 428/9), 1659 (undated).

(2) If the speeds are the same, but one body is bigger than the other, only the smaller will be reflected, and both will move on in the same direction with the common speed before Impact.

(3) If the bodies are equal but the speeds are different and in opposite directions, the slower will be carried away, and the common speed will be the mean of their initial speeds.

(4) If one is at rest, and a smaller strike it, that one will be reflected without communicating any motion to the stationary one.

(5) If a body at rest is struck by a larger one it will be carried away in the same direction with a speed which will be to that of the striking body as the mass of the latter is to the combined mass. A body having unit mass, being struck by one having double the mass, both will move together with a speed two-thirds the initial speed of the striking body.

In this last rule Descartes had the good fortune to light upon the truth. His cardinal error lay not so much in his assumption that the quantity of motion was unalterable, but in his failure to recognize that this " quantity of motion " had a definite sign, positive or negative, according to the direction in which it was considered. It remained for Wallis to show that what was unaltered by impact was the sum of the momenta measured in the same direction. It is therefore no injustice to Descartes to assert, as does Whewell, that the *Principia* of Descartes did little for physical science of the period. "If we were to compare Descartes with Galileo ", says that distinguished historian, " we might say that of the mechanical truths which were easily attainable in the beginning of the seventeenth century, Galileo took hold of as many, and Descartes of as few, as was well possible for a man of genius " *.

Such, however, was the poverty of the mechanical knowledge of the period, that despite the struggles of Stevinus and Galilei, Descartes' errors found ready acceptance amongst many of his contemporaries and successors. Fabri's *Dialogi Physici* (1665) merely repeated the blunders committed by Descartes ; Borelli's *De Vi Percussionis* (1667), though described by Wallis as work of considerable merit, was little better. Like so many contemporary treatises, it lacked sufficiently precise ideas concerning fundamental

* Whewell, ii, p. 39.

CHAPTER VII

notions, and though its author had grasped the important truth that "velocity is, by its nature, uniform and perpetual" *, his ideas were far too vague to do much for the science of mechanics.

But although there was little that might endure in Descartes' mechanical investigations, nevertheless they were indirectly responsible for one valuable result at least. Roused to a sense of the growing importance of mechanics, yet mindful of the blemishes in Descartes' work, the Royal Society, in 1668, initiated an investigation into the laws relating to the collision of bodies, and this proved a great stimulus to the more mechanically minded of the scholars of the period. In response to this invitation three geometers attempted an elucidation of the problems raised, and each communicated his results to the Society—Wallis on November 15th, 1668, Wren on December 17th, 1668, and Huygens, who was abroad at the time, submitted his observations early in the next year. From the *Tanner Papers* in the Bodleian we learn that Oldenburg also invited James Gregory " to examine the laws of motion given by Descartes, (a very ingenious mathematician, looking upon some of them as false) and having done that you would declare to us your meditations on that important subject " †. Wallis in his paper dealt almost exclusively with bodies absolutely hard (*i. e.*, bodies whose elasticity is zero), whilst Wren and Huygens discussed the impact of elastic bodies.

Long before this enquiry, however, Wallis had been conscious of the poverty of the knowledge of the laws governing mechanics, and more than once, by virtue of his position as President of the Oxford Philosophical Society, and his consequent voluminous correspondence with the Society in London, he strove to stimulate interest in the subject. He realized, as apparently did few others, its fundamental importance, and his letters abundantly illustrate this. As an example we may quote from a letter to his friend Boyle : " How much the World and the great Bodies therein, are managed according to the *Laws of Motion*, and *Statick Principles*, and with how much more of clearnesse and satisfaction, many of the more abstruse *Phænomena* have been salved on such Principles, within this last century of Years, than formerly they had been ; I need not discourse to you, who are so well versed in it. For, since that Galilæo, and (after him) Torricellio, and others have applyed *Mechanick* Principles to the salving of *Philosophical* Difficulties ; *Natural Philosophy* is well known to have been rendered more intelligible, and to have made a much greater progresse in lesse than an hundred years, than before for many ages " ‡. Not

* Quoted by Whewell, *op. cit.*, p. 20.
† *Tanner Papers* (1668–71), xliv, fol. 75, Jan. 19, 1668/9 (Bodleian).
‡ Wallis to Boyle, April 25, 1666. (*Royal Society Guard Books*.)

only that, but by his frequent insistence that its laws should be subjected to rigorous experimental scrutiny, he rendered inestimable service to the science of mechanics. " I only take for granted what sense and experience shews " *, he wrote to Oldenburg. On another occasion, in order to explain the wayward behaviour of the quicksilver " well purged of air " in the Torricellian Tube, he wrote: " Experiment will be the best judge " †, and in the same letter he gave details of a large number of carefully planned experiments which might with considerable advantage be tried. In an age in which scholars seemed to find satisfaction in *à priori* reasoning rather than in an appeal to ordinary common observation, the value of this aspect of Wallis's work cannot be overestimated.

Moreover, Wallis seems to have been particularly alert in grasping the true significance of the Law of Inertia, which had already been imperfectly delivered by Galilei and hinted at by others. " A Body in motion will continue in motion and with the same celerity, and the same way, till some positive cause do either stop or alter it " ‡, he wrote to Oldenburg in 1666, and three years later he wrote: " A positive cause is equally necessary and in the same degree of strength to stop a Motion as to begin it. Nor will a Body (once in motion) come to rest itself than (being at Rest) will move of itself " §. But how difficult was the formation of correct dynamical ideas can hardly perhaps be better illustrated than by the fact that the conception of what was afterwards called Inertia was beyond the ken even of Neile, for he postulated that " the Vis Motrix which now moves one Pound, so soon as it incounters a body at rest of 99 pounds must carry both, that is, 100 pounds with the same speed " ‖. To which Wallis replied: " For body at rest hath a repugnance to motion, and body in motion hath a repugnance to rest, though Body as Body is indifferent to either, and will therefore continue as it is (whether at rest or in motion) till some positive cause alter its condition. And when such positive cause comes, it acts proportionally to its strength, the lesse the strength with which it moves, and the heavier the body to be moved, the slower will be the motion " ¶. But Neile remained unconvinced, as is evident from others letters of Wallis, and the fact that a scholar of Neile's sagacity should have persisted in an error of such magnitude is not easy to understand. Feeble as had been Descartes' contributions, they had never been so completely opposed to common experience as was this.

* Wallis to Oldenburg, Jan. 2, 1668/9. (*Royal Society Guard Books.*)
† *Ibid.*, Sept. 26, 1672.
‡ *Ibid.*, March 21, 1666/7.
§ *Ibid.*, May 10, 1669.
‖ *Ibid.*, July 29, 1669.
¶ *Ibid.*

Wallis had long been interested in Gravity, a subject which after centuries of neglect was at last beginning to attract the notice of men of science. The conception of Gravity as a uniform force had always presented difficulties. Descartes could not accept it. " It is certain ", he wrote in a letter to Mersenne, " that a stone is not equally disposed to receive a new motion or increase of velocity when it is already moving very quickly, and when it is moving slowly " *. But the seeds which had been sown by Galilei half a century earlier had not altogether fallen upon stony ground. Interest in this subject was now thoroughly awakened ; indeed, nothing attests more clearly to the growing enthusiasm with which scholars began to pry into this fundamental concept than the fact that considerably less than one hundred years separate Galilei's investigations into the motion of falling bodies and Newton's mighty conception of Universal Gravitation. Such amazing progress speaks eloquently of the zeal with which the subject began to be prosecuted, and by no one was it more vigorously attacked than by Wallis, whose insatiable curiosity into the problems it presented is reflected again and again in his correspondence.

The theory governing the motion of projectiles, which had been elucidated by Galilei, was further prosecuted by Wallis, but unlike the speculations of all the earlier enquirers, save Galilei and Baliani, Wallis's results were backed by adequate experimental verification. He satisfied himself, for example, by severe trial, that " which way soever a heavy body be violently cast (upward, downward, horizontally or at any angle of inclination thereunto,) the naturale motion of Descent by reason of its gravity, (which with the motion of projection makes up the compound motion) is still the same " †, and he urged that this was the outcome of a rigorous examination by trial. Indeed, his observations on the motion of falling bodies, which he invariably tested by an appeal to experiment, constitute a fitting complement to the work of Galilei on the same subject, and, it may be added, an offset to many of the notions of Descartes which still found adherents.

Most of his researches on Gravity were elaborated in his work *De Motu*, and they will be described in due course. But at this point we might direct attention to an important paper, *De Gravitate et Gravitatione, Disquisitio Geometrica, Phoenomenis Experimento Comprobatis Stabilita* ‡, which was presented to the Royal Society on November 12th, 1674. Untrammelled by investigations into the ultimate " cause " of gravity (" quid sit, aut unde orta "),

* Descartes to Mersenne (Whewell, *op. cit.*, p. 26).
† Wallis to Sir Robert Moray, July 14, 1668. (*Royal Society Guard Books.*)
‡ " A Discourse on Gravity and Gravitation, grounded upon Experimental Results."

he made some penetrating observations on the subject. One passage is of peculiar interest : " Not only *Motus*, but *Conatus ad Motum* is properly Gravitation ", for underlying this is the concept of Potential Energy. The paper deals mainly with problems in hydrostatics, and with experiments on the newly invented Torricellian Tube, by which the doctrine of the counterpoise of the air is maintained and defended against that of the *Fuga Vacui* of the Ancients. " The Torricellian Experiment (with others of the same nature) is confessedly solved by the Pressure of the Air, which was anciently thought to be a *Fuga Vacui*. For if the air be heavy, it must gravitate, that is endeavour a Descent, (as other Heavy Bodies do) and actually effect it, if not opposed by at least as great a strength. And the Spring of the Air (allowing it to have a Spring) must always be of such a texture, as is Equivalent to the Weight or Force which it bears." Yet even a scholar of Wallis's ability could be misled occasionally, and in his anxiety to jettison the doctrine of the *Fuga Vacui* we find him going to the other extreme in explaining by means of the newly discovered pressure of the air phenomena with which it was only remotely connected. " It is observable also that water in very slender Pipes will rise visibly higher than the surface of that in the broad Vessel : because the air can more conveniently apply its pressure on that broader Vessel than in the slender Pipe." But when he determines the " different celerities of water's effusion " at various depths below the surface, as he does in one of his letters to the Society*, we recognize a distinctly erudite contribution to hydrostatics, and one which, like most of Wallis's investigations from first principles, is amazingly thorough and painstaking.

To return to Wallis's attempts to elucidate the laws of Impact of Bodies. The results of the investigation which he undertook at the invitation of the Royal Society are contained in the communication already referred to (November 15th, 1668). This is extended and elaborated in his treatise *De Motu*, which was published two years later.

In discussing Impact, Wallis observes that bodies fall into three classes. Bodies which suffer no distortion upon impact are termed absolutely *hard*. Those which are endowed with the faculty of altering their shape upon impact, and immediately recovering it, are said to be elastic, whilst the third class, *soft* bodies, suffer distortion and do not recover their shape. Here a part of the force of percussion is used up in deforming the body. Wallis does not consider these further ; his investigation begins with hard bodies. Later (Cap. xiii, *De Motu*) he extends the enquiry to bodies perfectly elastic.

* Wallis to Oldenburg, October 24, 1687. (Add MS. D. 105. Bodleian.)

CHAPTER VII

As a starting-point for the investigation he enunciates the theorem : If an agent A produce the effect E, then the Agent 2A will produce the effect 2E ; 3A, 3E etc., and universally, mA will produce the effect mE, putting m for the " exponent " of any ratio *. This is easily extended to the theorem :—If the power of Force (Vis) V can move a weight (pondus) P with a celerity C, then the force mV will either move the weight P with a celerity mC, or the weight mP with the same celerity, or lastly, any weight with such a celerity that the product of weight and celerity is mPC. It will be seen that the controlling factor in Impact is *Vis*, or what we call *Momentum* (Momentum, seu Vis Impellens, *De Motu*, Pars Tertia, Cap. xi, Prop. 2), *i. e.*, the product of weight (pondus) into the velocity. By this *momentum* the force of percussion is to be determined, and this seems to be one of the earliest attempts to measure the magnitude of a force by the effect it produces. Moreover, Wallis tacitly assumes the equality of action and reaction ; a body struck destroys in the striking body as much motion as it receives. " In all percussion the body striking looseth of its swiftness, and the other gains if before at rest ; If both before were in opposite motions, both loose of theirs, and both gain from the other in such proportion as is there expressed " †.

These principles being once established, Wallis enunciates the following theorem : " If a mass P, moved by a force V with a velocity C impinge directly against a quiescent mass mP (not obstructed) they will go on together with a celerity $\frac{1}{m+1} \cdot C$. For the greater the mass to be moved (the force being the same) so much less is the celerity ; namely, as

$$V : PC : : V : \frac{(1+m)P}{1} \times \frac{1}{(1+m)} \cdot C = PC."$$

Furthermore, if the mass P moved by a "force" V with celerity C be struck with another of mass mP moving with a celerity nC, (and therefore having a "force" mnPC, or mnV) then the common speed after impact will be

$$\frac{1+mn}{1+m} \cdot C$$

or

$$\frac{1-mn}{1+m} \cdot C$$

* " Si Agens ut A Efficit ut E, Agens ut 2A efficiet ut 2E ; 3A, ut 3E etc., ceteris paribus. Et universaliter mA ut mE cujuscunque Rationis exponens sit m." An account of ye Laws of Motion, sent to Mr. Oldenburg, Oxon, Nov. 15, 1668, and produced at the Society, Nov. 26, 1668. (*Royal Society Guard Books*.)

† Wallis to Oldenburg, Dec. 3, 1668. (*Royal Society Guard Books*.)

according as to whether the bodies are moving in the same or in opposite directions. These results of course only apply if the bodies impinge directly; in oblique impact the *Impetus* (*i. e.*, the product of the mass and the velocity) of the body striking is to that of its direct impetus " as the radius to the secant of the angle of obliquity ".

All this of course marks a tremendous advance upon the vague speculations of Descartes. There is still to be noticed a certain indefiniteness in the use of the words " force ", " impetus ", etc. This was perhaps understandable in view of the confusion of thought which had prevailed for so many centuries; apart from that, however, the two succeeding centuries have not added much to what Wallis contributed to the subject. In fact his letter to the Royal Society, which contains the results of these investigations, might not unjustly be regarded as one of the corner-stones of theoretical mechanics of the latter half of the seventeenth century; certainly Newton was under no misapprehensions as to its value when he referred to it in the Scholium to Corollary vi, subjoined to his *Leges Motus*.

According to the laws which Wallis had established in this paper, two bodies having equal momenta will, upon impact, reduce each other to rest. This, it will be recalled, was quite contrary to what Descartes had maintained; not only that, it seemed to be contradicted by experience, for Wallis's laws only applied to the purely hypothetical case, namely, when the striking bodies were perfectly hard. Therefore the conception of elasticity has to be introduced. " If bodies were perfectly hard ", he wrote to Oldenburg, " there would be no rebounding, and therefore it was necessary to have recourse to elasticity." This elasticity, or " springiness ", is the cause of rebounding. " If the bodies are elastic ", he continued, " yielding to the stroke, and then restoring themselves by an equal force, the bodies instead of moving together may recede from one another, and that more or less in proportion to the restoring force .... That all rebounding comes from springynesse, is my opinion. Quiescent matter has no resistance to motion (save what it may have from circumstantial encumbrances), or if there be any innate propensity to a contrary Motion, as in Gravity is supposed, I take for granted " *.

The views expressed by Wallis on the subject of Impact were so foreign to the notions then current that they were promptly challenged, and Wallis was called upon to defend them, using the principal means then available, namely, his private correspondence. Some of his letters are remarkable for the painstaking thoroughness

* Wallis to Oldenburg, Dec. 3, 1668. (*Royal Society Guard Books.*)

with which he maintains his views. Not only that, they also give an indication of Newton's indebtedness to Wallis, for many of the principles which form the groundwork of Newton's researches are clearly discernible in the writings of Wallis. In fact, it becomes more and more evident as we read through the pages of these letters that when Newton enunciated the famous laws which bear his name he was merely giving utterance to what had been in Wallis's mind for many years. In these letters, Wallis's cardinal points are :—

(1) " Quiescent matter is indifferent to rest or to motion, without any averseness to either, as also indifferent to any direction of motion, and accordingly it doth remain as it is either in rest or in motion and this with the same direction and celerity till some positive cause alter it."

(2) " In Percussion the force or Impetus whereby one body is moved may cause another body against which it strikes to be put in motion, and withal lose some of its strength or swiftness."

(3) " Different motions may destroy one another by making a compound motion of both, which may chance to be rest " *.

Two other aspects of Wallis's investigations in mechanics are at this point worthy of notice. One is the enunciation of a principle which is tacitly assumed in every modern treatise. " The hypothesis I sent ", he wrote to Oldenburg, " is indeed of the *Physical* Laws of Motion but *mathematically* demonstrated. For I do not take the Physical and Mathematical hypotheses to contradict one another at all. But what is physically performed is Mathematically measured. And there is no other way to determine the Physical Laws of Motion exactly but by applying the Mathematical measures and proportions to them " †. It was views such as these that helped to make mechanics one of the most exact sciences of succeeding centuries. Secondly, he is ready to accept such fundamental notions as Gravity, Motion, etc. as postulates, *i. e.*, these things *are* ; it is useless therefore to speculate as so many did, into their ultimate causes. " What that is which we call springynesse and what Gravity I do not determine, but from those things, whatsoever they are, and from whatever causes they proceed, I am to give account of the effects I ascribed to them. The one (from whatever cause) is the principle of the motion of restitution, and the other of the tendency downwards. I know Des-Cartes and others do attempt to assign causes to both, but I have not seen any hypothesis that doth fully satisfy my apprehensions " ‡. This was indeed a change

* Wallis to Oldenburg, Dec. 5, 1668.   (*Royal Society Guard Books.*)
† *Ibid.*
‡ Wallis to Oldenburg, Dec. 31, 1668.   (*Royal Society Guard Books.*)

in outlook. Wallis never allowed himself to be hampered by enquiries of this nature ; these were not the business of the man of science. Newton of course adopted a similar attitude, but there appears to be no doubt that one of the causes which militated against the development of the subject was the fact that for a long time there were not wanting those who thought that to probe into ultimate causes was of primary importance.

Meanwhile Wallis had gathered together the results of his erudite researches in mechanics. These were published in his celebrated treatise *De Motu*, the first part of which appeared in 1669, and the remaining two in the year following. The work was completed in 1670, and appeared in one volume under the title *Mechanica, sive De Motu, Tractatus Geometricus*. The treatise is described in the *Philosophical Transactions* for 1695 (no. 216, p. 74) thus : " His *Mechanica*, or a large Treatise of *Motion*, Wherein are handled, not only the Machines or Engines, commonly called the *Mathematical or Mechanical Powers*, but the whole Doctrine of Motion ; derived and demonstrated from its Genuine and first Principles ; with great variety . . . . of intricate and perplexed Enquiries into the most abstruse Mysteries in Mathematicks, many of them not formerly handled by any ; the Doctrine of Percussions, Repercussions, Springs and Reflexions ; the Doctrine of Hydrostaticks, from the Counterpoise of the Air, and many other things newly discovered ". But even this comprehensive catalogue does but scant justice to the vastness of the field over which Wallis ranged in this important work, a work which was without doubt the most exhaustive exposition of the subject that had so far appeared. Not undeservedly the work evoked the warmest praise ; the Oxford Editor of Oughtred's *Opuscula Mathematica hactenus inedita*, Oxon, 1677, for example, speaking of Wallis's treatise concludes with the elegant eulogy : " Whatever is required in mechanics can be adequately found in that most accurate treatise of Wallis, who alone hath treated that aspect of mathematics with dignity, and whosoever shall in future be pleased to write concerning this, must needs borrow from him " *. Gregory, in his *Life of Wallis*, to which we have already referred, speaks of the treatise as a " full and complete Mechanics in which he lays the true and solid foundations of the Science, which had not been done before " †.

The treatise contains an elaborate investigation into the determination of Centres of Gravity and into the nature of the different

* " In institutionibus mechanicis, si quid desideretur abunde suppleri potest ex accuratissimo tractatu Cl. Wallisii, qui solus pro dignitate hanc partem Matheseos tractavit, et de qua si cui postea scribere libuerit, ab illo mutuetur necesse est.

† Gregory, *Life of Wallis*. (Smith MS. 31. Bodleian.)

mechanical powers, and it is remarkable for the free use of the Principle of Virtual Displacements. The motion of projectiles is discussed, the effect of a resisting medium being considered, and problems on Percussion are more elaborately treated than ever before. The work, however, is of such consequence as to call for a more detailed account; we pass therefore to a review of this important treatise.

Wallis indicated his general plan in the Dedication to Part I, which preceded by nearly two years the complete work. Addressing Lord Brouncker the author declared : " You have now the First Part, which contains the fundamentals of the whole treatise, and particularly the Doctrine of the Balance. Soon after will follow the Second Part, which deals with Centres of Gravity and the calculation thereof in most curvilinear figures and the solids and the curved surfaces arising from these ". This first part contains three chapters, headed respectively :

(1) De Motu Generalia,

(2) De Gravium Descensu et Motuum Declivitate,

(3) De Libra.

Like the Elements of Euclid, upon which Wallis seems to have modelled his work, the *Mechanica* is prefaced by an elaborate array of definitions. Mindful of the defects which had vitiated the work of so many other writers on the same subject, Wallis, with characteristic thoroughness, endeavoured on the very threshold to place the fundamental notions on a sound basis. Remembering his heritage of mechanical learning we cannot doubt that, save in a few cases, he achieved considerable success, and it is this aspect of his work which contributes so enormously to the value of the treatise. Wallis appears to have been the first to realize the futility of discussing mechanical questions so long as the terms used had no definite and permanent meaning, and it cannot be too often asserted that not the least reason for the tardy progress which had been made in mechanics during the preceding century was the confusion of ideas on fundamental notions which prevailed. It is very interesting to note, too, that the dignity of the subject never escaped his notice. Mechanics was to him the Geometry of Motion ; it was therefore worthy of nothing less than the highest efforts of its votaries. This differed profoundly from the popular conception of mechanics, by which, says Wallis, it was regarded merely as the art of the craftsman demanding the use of hands rather than of intelligence. " The mechanical arts ", he complained in the first chapter, " are usually looked upon with contempt as the activities of menials and the like, which uncultured people engage in, and in the performance of which, toil seems to be needed rather than ingenuity. And they are usually distinguished not only from geometry, but even from those other activities which usually occupy the mind rather than the hands and which either demand or develop astuteness of intellect " *. This was a change in outlook, and one which was long overdue. It is certain that the " mechanic art " had been despised for centuries as being the work of menials in the slave order of Greek and Roman times, and it is more than likely that the atmosphere of improbity which seemed to cling so tenaciously to the subject was yet another reason for the neglect which it suffered through the centuries.

The order in which the definitions are arranged would, however, appear strange in a modern treatise. Traces of Scholasticism still persist, and his definition of *Motion* (*Per Motum intelligimus Motum localem*, Def. 2) is not so much a definition as a distinction from the notions of generation, augmentation, and the like which were part of the stock in trade of the logician of the Middle Ages.

Force (*Vim motricem, vel etiam Vim simpliciter appello*) is defined as that which is capable of causing motion—"*potentiam efficiendi motum* ", a definition not unlike that afterwards employed by Newton. But even Wallis was never very clear about the notion of force ; in a paper to the Royal Society † he translated *Vis* by " Force or Power ", and he uses both terms indiscriminately. One cannot help noticing how slow scholars were at grasping this fundamental concept. Most writers of the period, e. g., Baliani, when speaking of *Vis* usually meant *Impetus*, and as late as 1815 Hutton (*Mathematical Dictionary*) makes little distinction between Force, Momentum, and Impetus. Wallis, when he speaks of Force in dynamical problems, almost invariably means what we should call *Momentum*, i. e., the product of mass and velocity. Thus in the communication upon Impact already referred to it is clear that when he speaks of *Force* he is thinking of what we should refer to as *Momentum*. Again, he wrote to Oldenburg : " For the same *Vis* which carried A two spaces in a certain time ; when it finds another quiescent (and equall to it) will carry both, that is 2 A, but not above half so fast. For to carry 2A requires double the

---

* " *Mechanicæ artes*, per contemptum dici solent illiberales illæ Cerdonum artes, et his similium, quas rude vulgus exercet ; ad quas Labore magis quam Ingenio videtur opus. Et distingui solent, non à Geometria tantum, sed ab aliis etiam *Ingenuis*, (quæ mente magis quam manu exerceri solent ; atque acumen animi vel postulant vel faciunt." *De Motu*, Cap. 1. (*Opera* I, p. 575.)

† *Phil. Trans.*, no. 43, p. 864 (1669). (Hutton's Abridged Phil. Trans., vol. i, p. 308.)

strength of what carried A at that rate " *. In Proposition 27 he is more definite: " In comparatis Motibus, Virium gradus (cæteris paribus) sunt in ratione quæ ex Ponderum et Celeritatum rationibus componitur ", i. e., in comparing motions, the " degrees " of the forces are as the products of the weights and the velocities.

This conception of *Force*, namely Pondus (or Mass) multiplied by Velocity, is what Descartes had used to express the " quantity of motion in a body ", and it is probably what was generally understood by the " force of a moving body ". Wallis occasionally lapsed, and spoke of this product as the *Impetus*; e. g., in the paper to the Royal Society which we have mentioned he wrote : " The Impetus of one body, (that is the product of the mass and the celerity) . . . ." Nevertheless he had clearly grasped the importance of this conception; on it, he says, depends the solution of the most important problems in mechanics, as, for example, the principle that " the magnitude of the mass compensates for the slowness of the motion " †, and again, " In whatever ratio the weight is augmented, in the very same ratio is the celerity diminished ".

Nevertheless the word *Momentum* is introduced in Definition 3 " *Momentum appello id quod motui efficiendo conducit* ", i. e., " I call Momentum that which tends to the production of Motion ", whilst its opposite, namely, that which opposes motion (*id quod motui obstat, vel eum impedit*), is called *Impedimentum*. This definition, which is reminiscent of the " efficient causes " of the Scholastics, is, however, not very helpful. But the word had already been overworked. The term Momentum had been introduced to express the force of moving bodies long before it was even surmized what its effect was. Galilei, in his *Discorso intorno alle Cose che stanno in su l'Acqua*, says : " Momentum is the force, efficacy, or virtue with which the motion moves and the body moved resists, depending not upon weight only, but upon the velocity, inclination and any other cause of such virtue " ‡. Later he realized that Momentum was proportional to mass and velocity jointly, though often he spoke of this same product as the *Impulse*, and sometimes as the *Energy*. Wallis used *Momentum* where we should *Force*, as in the definition quoted above, and in Proposition 11, where he states : " *Si Momentum Impedimento præpollet*; *Motum efficit* ", i. e., if the momentum exceeds the resistance motion results. But in Proposition 17 he makes momentum *proportional* to the force, and in Proposition 19 to the time ; in Proposition 21 it is proportional to the two combined : " *Si Vires et Tempora sint vel utraque æqualia vel sint reciproce proportionalia, quæ hinc resultant Momenta sunt*

* Wallis to Oldenburg, May 17, 1669. (*Royal Society Guard Books*.)
† *De Motu*, Scholia, Prop. 27, 28. (*Opera* I, p. 593.)
‡ Whewell, *op. cit.*, ii, 38, 39.

æqualia ", i. e., if the forces and the times are equal, each to each, or reciprocally proportional, the resulting momenta are equal. Thus his momentum becomes our *Impulse*. In Chapter iii, when he discusses statical problems, he uses the same word *Momentum* for the product of force into distance, i. e., the modern *Moment of a Force about a Given Point*, although, as if conscious that the word had already been given another significance, he smuggles in another word *Ponderatio* (*cum enim Ponderationis sive Momenti ratio, ex rationibus Ponderum et Distantiarum componatur*—Proposition 15). This was a good word ; it is unfortunate that it was allowed to fall into desuetude.

Gravity is defined (Definition 12) merely as *Force Downwards* (*vis motrix deorsum*). Its measure is *Pondus*. " *Per Pondus intelligo gravitatis mensuram* " (Def. 13). This definition recalls that given by Stevinus : " *La pesanteur d'un corps, c'est la puissance qu'il a de descendre, au lieu proposé* " *.

Wallis was well aware that the ideas concerning weight were in a very confused state, and he tried to disentangle them. He had only the vaguest notion of the difference between Mass and Weight, and he used the word *Pondus* almost indiscriminately for both. Half a century earlier Baliani had described an experiment on the rate of fall of a ball of iron and one of wax, which led him to distinguish between *moles* and *pondus*, the former being somewhat akin to Newton's idea of *Mass* or amount of matter. Wallis used the two terms *Pondus* and *Onus*. " *Ego autem, neglecto si quod est inter Pondus et Onus discrimine, (quo Libram, illud ; hoc, Vectem magis spectet) per Pondus jam intelligo, illam, in utrovis, gravitatis mensuram, quam ad Libram solemus examinare* " †, which suggests that *Pondus* meant *weight* to Wallis. But more frequently he uses the term where a modern writer would use *Mass*, e. g., " *Ponderis nomine, plerumque, Vim Resistentiæ in sequentibus designabimus et Virium nomine vim Motricem* "—that is, by the name *Pondus* is to be meant for the most part the force of resistance. Again he wrote : " *Pondus, sic intellectum, aut Gravitas etiam ; prout vel in Movente, vel in Mobili, consideratur ; ita vel ad Movendi ; vel ad Resistendi, vim pertinebit. Adeoque nunc ad Momentum, nunc ad Impedimentum referetur* ". (*Pondus* thus understood is considered either in a moving body or in a movable one, and thus it may relate to the force of moving or to the force of resistance. Thus now it refers to *Momentum*, now to *Resistance*.)

* Stevinus, *Œuvres* (1634). Livre de la Statique, Def. 2, p. 434.

† " Ignoring any difference that there may be between *Pondus* and *Onus* (the former relating to the Balance, the latter to the Lever) by the name *Pondus* I understand that measure of Gravity which we are accustomed to determine by the Balance." Def. 12. (*Opera* I, p. 577.)

CHAPTER VII

Having resolved to treat his subject mathematically, Wallis devotes the first few propositions to preparing the mathematical groundwork. Elementary propositions on ratios, such as the constancy of a ratio when antecedent and consequent are multiplied or divided by the same factor, are proved *in vestibulo* on account of their frequent recurrence.

In Proposition 11 the first Law of Motion, the truth of which had often been implied, appears in a dress only slightly different from that in which it appeared nearly twenty years later. " *Si Momentum Impedimento præpollet, Motum efficit; Adeoque; si nullus fuerit, Inchoatur : Si jam fuerit, Augetur.*

" *Si præpollet Impedimentum : Impedit. Adeoque Motum, siquis jam sit, vel Tollit, vel saltem Minuit .... Si Æquipollent : Neque ponitur Motus, neque Tollitur. Adeoque quæ prius erat, vel Quies vel Motus perseverat ".*

(" If the force is greater than the resistance, motion will result. If there was none before, motion will begin ; if it already existed, it will be increased. If the resisting force is greater, it will oppose the motion if there was any, or at least check it. If the two are equal, motion will be neither started nor stopped, and the initial state of the body, either of rest or of motion, will persist.")
In Proposition 12 almost the same statement occurs except that Force (Vis) is substituted for Momentum : " *Vis vi contraria, si æquipollet sustinebit: Si minus pollet; ne hoc quidem .... Si præpollet, movebit* ", i. e., a force will neutralize an equal and opposite force ; if it is less, it will not even do that ; if it is greater it will move. Nevertheless, although Wallis was hazy regarding the true conception of *Force*, and although he seemed to use the terms Force, Momentum, Impetus indiscriminately, he had really grasped their interrelation ; for in Propositions 17 to 21 he explains that the momenta of forces are proportional to the forces and to the times during which they act. If these forces and times are equal or reciprocally proportional, the resulting Momenta are equal. Now if, as we have seen, his *Momentum* is our *Impulse*, and his *Force* our *Momentum*, Wallis's statement becomes almost identical with the Second Law of Motion, namely, that the Momentum produced is proportional to the Impulse of the Force. This was not the only time that Wallis realized the proportionality between the force and the motion produced by it ; indeed, he seems to have been thoroughly familiar with the notion many years before it was given such definiteness by Newton in 1687.

Having established the above general principles, Wallis opens up in Chapter ii an investigation into the subject of the motion of bodies under the action of gravity, *De Gravium Descensu, et Motuum Declivitate*. This begins with the Proposition : " Heavy bodies, other things being equal, gravitate in proportion to their weights " *. This is a particular application of a general principle in which he placed great reliance, namely, that any forces (*Vires Motrices*) produce effects proportional to their magnitudes. In this case gravity is the force, and its measure is *Pondus*; therefore gravity will cause a body to move in proportion to its weight.

This statement was never pursued by Wallis to its conclusion, namely, that since the weight of the body is the force which gives it its motion, and since as Galilei had shown all bodies irrespective of their weights move similarly under the action of their weights, *i. e.*, move with the same *acceleration*, there must be some property of a body which, though not identical with weight, is nevertheless proportional to it, namely, its *Mass*. Had Wallis been able to take this leap forward he would have anticipated much of Newton's most erudite work in this sphere.

In Proposition 5, too, there appears a familiarity with the conception of *Energy*. " *Gravium Descensus, invicem comparati, in ea ratione pollent, quæ ex Ponderum ratione et ratione Altitudinum Descensuum componitur* "—that is, the " force " of a falling body is proportional to the product of the mass and the distance through which it falls. Further, he realized that this principle was at the basis of such machines as the Inclined Plane etc. But for a clear enunciation of the Principle of the Conservation of Energy we have to wait for the great French mathematicians who succeeded Newton.

Chapter iii, *De Libra* (The Balance), deals with the first of the mechanical machines (*instrumenta motibus examinandis, vel etiam facilitandis, forinsecus adhibita*, Cap. i, Def. 22). With customary thoroughness the different parts are described, and according to the *Philosophical Transactions* (1669) the author " doth from their proper Principles demonstrate many of these things which writers commonly postulate or take for granted, but which (to make a sure Foundation) ought to have been demonstrated " †. The fundamental principle is enunciated in Proposition 12—the weights act in the ratio of their magnitudes and their distances from the centre of motion combined, and the equality of this ratio must persist for equilibrium, even if the beam of the balance be not horizontal. His investigation is very comprehensive, and Wallis's partiality for general solutions, which we have noted in the *Arithmetica Infinitorum* and elsewhere, is well marked. " It is seen ", he says, in the Scholium to Prop 14, " that in this Proposition which is enunciated in general terms, that many cases are considered and understood, which is better than demonstrating many propositions individually, for in the first place they all rest on the same

* " Gravia, cæteris paribus, gravitant in ratione Ponderum."
† *Phil. Trans.* (1669), p. 1089.

principle, they are all verified by the same demonstration, and it would be a very great labour to consider them individually, when they might all be considered as one ".

For what we call the Moment of a Force about a given point Wallis introduced the word *Ponderatio*, a word certainly more suggestive of the idea underlying it than our *Moment*. Lagrange, it may be noted, was of the opinion that Wallis was one of the few who had really grasped the idea of a Moment. " Il me semble ", he says in his *Mécanique Analytique*, " que la notion du *Moment* donnée par Wallis et par Galilée est bien plus naturelle et plus générale, et je ne vois pas pourquoi on l'a abandonée pour y en substituer une autre qui exprime seulement la valeur du moment dans certains cas, comme dans le levier " *. This *ponderatio* is the product of the force into the distance. Wallis uses both terms, *ponderatio* and *moment*, e. g. " *Momentum illud hic intelligo, quo Pondus gravat suum respective Libræ Brachium ; quod itaque speciatim Ponderationem appello* " †. This *ponderatio* vanishes about the line of action of the force : " *Et, nullius Distantiæ, nulla est Ponderatio, quodcunque sit Pondus* ", i. e., if the distance be zero, the ponderatio will also be zero, no matter how great the force.

The latter part of the chapter is taken up with the investigation of the conditions of equilibrium of a beam loaded at different points, *e. g.* (Proposition 18) in figure 12 :

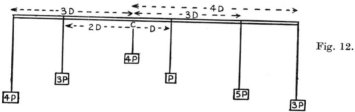

Fig. 12.

Weights of 4P, 3P act at distances 3D, 2D respectively to the left of C, the point of support ; weights of P, 5P, 3P act at distances D, 3D, 4D to the right. A weight 4P acts at C, but its " ponderatio " is 4×0, *i. e.*, 0 (P is merely his unit of weight, as D is that of distance).

Now the product $P \times D$ is indicated by G ; the weights on the left, therefore, " ponderant " $12G + 6G = 18G$, whilst those on the right have a total *ponderatio* equal to $G + 15G + 12G$, or 28G. Hence to balance the beam there must be a *ponderatio* of 10G on the left, *e. g.*, a weight of 10P at a distance D from C.

Similar results lead to the generalization in Proposition 20.

* Lagrange, *Mécanique Analytique* (Paris, 1788), p. 9.
† Scholium, Prop. 15. (*Opera*, I, p. 632.)

"Weights $r\text{P}$, $s\text{P}$, $t\text{P}$, etc., hung at distances $+l\text{D}$, $+m\text{D}$, $-n\text{D}$ ... will have Moments $+lr\text{DP}$, $+ms\text{DP}$, $-nt\text{DP}$, etc. If the aggregate of all these be divided by the sum of the weights we get the distance

$$\frac{+lr\text{DP}+ms\text{DP}-nt\text{DP}}{r\text{P}+s\text{P}+t\text{P}}$$
$$=\frac{lr+ms-nt}{r+s+t}\cdot\text{D}$$

(right or left according to whether the sign $+$ or $-$ prevails), at which, if the sum of the weights was hung, it would have the same moment "*.

Important though this investigation was, there was probably little in it that was new. But Wallis's demonstrations were immeasurably more exhaustive and thorough than anything which had gone before. It was essential that they should be, for this investigation on the Principle of Moments and its application was the prelude to a much more important piece of research, namely, an investigation into the methods of determining the Centre of Gravity of different figures, which, after all, is nothing more than the Principle of Moments in its generalized form. This enquiry begins at Chapter iv, *De Centro Gravitatis*. On the threshold are recounted the familiar properties of the Centre of Gravity which had already been elaborated by Galilei and Stevinus, namely, a body supported at its Centre of Gravity is in equilibrium; there is only one Centre of Gravity, and this lies on any axis of equilibrium; the Centre of Gravity of two bodies joined together lies on the line joining the two Centres of Gravity.

Wallis's method of finding the Centre of Gravity is based upon three principles :—

(1) Every body (or *continuum*) can be considered as being made up of a very large number of parts. " Continuum quodvis (secundum Cavallerii Geometriam Indivisibilium) intelligitur, ex Indivisibilibus numero infinitis constare " †.

(2) The weights of these parts are proportional to their magnitudes.

(3) If the sum of all these particles balances about any axis, the Centre of Gravity must be taken on that axis.

* " In distantiis $+l\text{D}$, $+m\text{D}$, $-n\text{D}$ etc. appensa pondera $r\text{P}$, $s\text{P}$, $t\text{P}$, etc. ponderant ut $+lr\text{DP}$, $+ms\text{DP}$, $-nt\text{DP}$ etc. Horum aggregatum, si per summam Ponderum dividatur; quod prodit

$$\frac{+\ lr\text{DP}+ms\text{DP}-nt\text{DP}}{r\text{P}+s\text{P}+t\text{P}}=\frac{lr+ms-nt}{r+s+t}\cdot\text{D}$$

est Distantia (Dextrorsum aut Sinistrorsum, prout notata signis $+$ aut $-$ præpollent) qua si summa Ponderum suspendatur, similiter Ponderabunt." Prop. 20. (*Opera*, I, p. 637.)

† *De Motu*. (*Opera* I, p. 645.)

CHAPTER VII

Now if a body be divided into a large number of small parts, not necessarily equal but at least increasing or decreasing according to some known law (e. g., a series of *Æquales, Primana, Secundana,* etc.) *, their total sum may be ascertained by the methods already elaborated in the *Arithmetica Infinitorum*. Hence Wallis begins (chapter v, *De Calculo Centri Gravitatis*) by revising some of the fundamental propositions of this earlier work, culminating in the generalized theorem that

$$\frac{0^n+1^n+2^n+3^n+\ldots+s^n}{s^n+s^n+s^n+s^n+\ldots+s^n} = \frac{1}{n+1}.$$

There is a marked improvement in notation; indeed, apart from a persistent haziness concerning Infinity the work is not materially different from a modern treatise. But he still adheres to his belief in quantities " greater than Infinity ", which is strange, because very often his conception of Infinity is on distinctly modern lines.

The whole investigation is nothing more than an extended application of the Law of Moments. Masses lP, mP, nP, . . . at distances rD, sD, tD, ... from any axis will have moments about that axis lrDP, msDP, ntDP . . . and the sum of these (by Proposition 17) is equal to the product of the total mass and the distance of its centre of gravity from the same axis. The only difficulty, therefore, lies in finding the sum of the masses lP, mP, nP . . . . which make up the whole figure.

Now it is easy to ascertain from elementary geometry the centre of gravity of regular plane figures such as parallelograms, etc., which can be divided into strips by equidistant lines drawn parallel to a side, or of circles and ellipses which can be divided into similar strips parallel to two perpendicular diameters. This is extended in Proposition 6; whatever figure can be divided into strips by lines drawn parallel to a side or an axis, and whose magnitudes beginning from 0 proceed according to a series of *Primana, Secundana*, etc., e. g., according to the series

$$0^n+1^n+2^n+3^n+\ldots+s^n$$

(where n may assume any value, including zero),

the magnitude of such a figure is easily ascertained by the principles already established in the *Arithmetica Infinitorum*, where it is shown that the area of such a figure bears to that of the circumscribing

* Def. 1. "Quanta quælibet, Arithmetice proportionalia, (sive secundum naturalem Numerorum consecutionem constituta) appello *Primana* : Quæque sunt in horum ratione Duplicata, Triplicata . . . . appello *Secundana, Tertiana* etc.", *i. e.*, quantities arithmetically proportional, or proceeding according to the natural numbers, I call *Primana*, those quantities which proceed according to the squares cubes, of these, I call *Secundana, Tertiana*, etc. Cap. v, p. 665. (*Opera*, I.)

rectangle the ratio $1 : n+1$. This opens up the way to the determination of the centre of gravity of a plane figure bounded by a curve whose ordinate is always some power of the abscissae, *i. e.*, a curve represented by the relation $y = x^n$.

Thus, suppose it is required to find the centre of gravity of a figure shown (fig. 13):

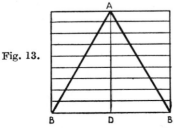
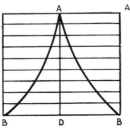

Fig. 13.

Whatever the series of small parts into which the figure is divided by lines drawn parallel to BB, *e. g.*, whether a series of

Æquales.	$0^0 + 1^0 + 2^0 + 3^0 + \ldots\ldots\ldots$	(Index$=0$).
Primana.	$0^1 + 1^1 + 2^1 + 3^1 + \ldots\ldots\ldots$	(Index$=1$).
Secundana.	$0^2 + 1^2 + 2^2 + 3^2 + \ldots\ldots\ldots$	(Index$=2$).
Tertiana.	$0^3 + 1^3 + 2^3 + 3^3 + \ldots\ldots\ldots$	(Index$=3$).
Subsecundana.	$\sqrt{0} + \sqrt{1} + \sqrt{2} + \sqrt{3} + \ldots\ldots$	(Index$=\tfrac{1}{2}$).
or generally,	$0^n + 1^n + 2^n + 3^n + \ldots\ldots\ldots$	(Index$=n$).

—if these are at distances 0, 1, 2, 3, ... (numbers arithmetically proportional) from the axis through A, then the several moments of these small parts about that axis will form a series whose index is one step higher than the index of the series of small parts. Moreover, since the sum of these moments is equal to the magnitude of the whole figure multiplied by the distance of its centre of gravity from the same axis, the determination of the latter is easily effected.

For example, in the triangle, the strips drawn parallel to an axis form a series $0p$, $1p$, $2p$, $3p$, ... P (fig. 14),

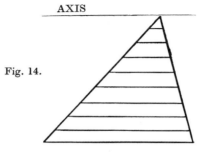

Fig. 14.

i. e., a series of *Primana*, whose index is 1. Hence the sum of n

CHAPTER VII

such strips is $\frac{1}{2}n\text{P}$. But if their several distances from the axis are

$$0d, 1d, 2d, 3d, \ldots \text{D},$$

their several moments about this axis will be

$$0dp, 1dp, 4dp, 9dp, \ldots \text{DP},$$

i. e., a series of *Secundana*, whose index is 2, and thus the sum of all these (by Proposition 19, *Arith. Infin.*) is equal to $\frac{1}{3}n\text{DP}$. Therefore equating moments, and remembering that the magnitude of all the strips is $\frac{1}{2}n\text{P}$, we get

$$\tfrac{1}{2}n\text{P} \times \text{distance of C.G. from axis} = \tfrac{1}{3}n\text{PD},$$

whence the required distance $=\frac{2}{3}\text{D}$, *i. e.*, the centre of gravity is two-thirds of the distance from the axis to the opposite side.

Similarly for the cone:

The strips are $0p$, $1p$, $4p$, $9p$,   ... P (sum $=\frac{1}{3}n\text{P}$).
Their distances are $0d$, $1d$, $2d$, $3d$,   ... D.
Their Moments are $0dp$, $1dp$, $8dp$, $27dp$,   DP.

i. e., a series of *Tertiana*, whose sum is $\frac{1}{4}n\text{DP}$, whence, as before,

$$\tfrac{1}{4}n\text{DP} = \tfrac{1}{3}n\text{P} \times \text{distance of C.G. from axis},$$

which gives as the required distance $\frac{3}{4}\text{D}$.

This important discovery is at once generalized, as was Wallis's custom, in Proposition 6, which may be enunciated as follows:—

"If the figure can be divided into a large number of parts whose magnitudes starting from a given axis form a series

$$0^n, 1^n, 2^n, 3^n, 4^n, \ldots \text{P},$$

and if there are N terms of the series, and the last (*i. e.*, the greatest) is P, then by Proposition 44 (*Arith. Infin.*) the sum of all these will be $\dfrac{1}{n+1} \cdot \text{NP}$; and, furthermore, the moments of these about the given axis will form the series

$$0^{n+1}, 1^{n+1}, 2^{n+1}, 3^{n+1}, 4^{n+1},$$

whose sum is $\dfrac{1}{n+2} \cdot \text{NP}$. So if this be divided by the sum of all the parts which make up the figure $\left(\dfrac{1}{n+1} \cdot \text{NP}\right)$, the quotient will give the distance of the centre of gravity from that axis; *i. e.*, the centre of gravity will divide the distance between the axis and the base in the ratio $n+1 : n+2$."

But when n was negative and greater than unity Wallis was again confronted with his earlier difficulty, namely, that of assigning a meaning to a ratio one term of which was positive and the other negative. It will be recalled that in determining the areas under curves which reduced to a ratio of this nature he had concluded that such areas were *greater than infinity*. In the case of the ratio here considered, however, namely $\frac{n+1}{n+2}$, since this was to him meaningless if n were negative and greater than 1, he concluded that such figures had no centre of gravity! ("Si sit index minor quam -1, puta -2, -3, etc. . . . Centrum Gravitatis non habent.") *.

The investigation is directed to all kinds of figures. In Proposition 14 curved figures are introduced, and these are followed by solids of revolution. At Proposition 20 the cycloid is introduced. He investigates many of the properties of the curve, and he shows that its length is four times that of the axis. This rectification had, of course, been accomplished a dozen years earlier, but Wallis broke new ground by examining many of its properties, by finding the lengths of different portions of the curve, and by determining its centre of gravity. It will be recalled that the rectification of the cycloid and the investigation of its properties had been the occasions of his clashes with Fermat and with Pascal, and although many years had elapsed since those unhappy affairs, and both Fermat and Pascal were dead, it is clear that the dispute had lost none of its bitterness. For not only does Wallis declare that his methods were as good as the Frenchmen's, he repeats the remark, as unjust as it was uncivil: "Quæ quidem si negligimus, Insultant; si solvimus; Irascuntur, calumniantur, opprobriis onerant, insimulant plagii" †. After this interlude Wallis passes to the spiral, the cissoid, and the conchoid, each of which is treated with a thoroughness hitherto unknown in the history of the subject. With the conchoid the chapter is concluded.

The third part of the *Mechanica* deals with the Balance and with the machines for facilitating work. There is also a chapter on the *Composition of Motions* (*De Motibus Compositis, Acceleratis Retardatis et Projectorum*), to which we shall refer later; another on Hydrostatics and the Pressure of the Air, and a concluding chapter: *Variis Quæstionibus Mechanicis*.

But probably the most original and at the same time the most stimulating chapters were *De Percussione* and *De Elatere et Resilitione seu Reflexione*, where problems on percussion are investigated

* *De Motu*, Cap. v, Prop. 7.
† "If however we ignore these, they insult; if we solve them they are angry; they calumniate us, load us with insults and accuse us of plagiarism." (*Opera* I, p. 860.)

CHAPTER VII

with marked thoroughness. Without doubt this part of the work had a pronounced influence upon the course of mechanical learning during the next fifty years or more. It will be remembered that Impact had been the subject of Wallis's noted communication to the Royal Society in the November of 1668. In this chapter he extends his investigation, dealing first with bodies inelastic and then with bodies elastic.

If a perfectly hard body impinge upon another, the subsequent motion of each is governed by their respective momenta before impact. Basing his investigation upon the Principle of Momentum, he shows how to find the common speed after impact—*si Momentum* (*ex gravis Moti Pondere et Celeritate compositum*) *per utriusque simul Pondus dividatur*; *habebitur futura Celeritas*, i. e., the resultant speed is obtained by dividing the combined momentum by the combined mass (Prop. 2). Thus if a body of mass mP and velocity rC impinge upon a body of mass nP at rest, both will move on together with a common velocity $\frac{mrPC}{mP+nP}$, i. e. $\frac{mr \cdot C}{m+n}$. The momentum lost by the striking body is $mPrC - \frac{mrC}{m+n} \cdot mP$, which is equal to $\frac{nPmrC}{m+n}$, and this is equal to the momentum gained by the stationary body $\left(nP \times \frac{mrC}{m+n}\right)$. Hence there is no loss of Momentum as a result of Impact. In these observations Wallis was beginning to approach the Third Law of Motion in its most generalized form, namely, that momentum may be taken for a measure of its effect, so that momentum is as much diminished in a striking body by the resistance it experiences as it is increased in a body struck by the impact, and the quantity of motion remains unaltered. From Wallis's observations it follows also that even the largest body can be set in motion by the impact of a body be it ever so small. "*Manifestum hinc est, ex quovis minimo cujusvis exigui Gravis impulsu, etiam maximo cuivis quiescenti, motum inferri posse*" *.

If both bodies are moving, namely a mass mP with speed rC and a body of mass nP and speed sC, then the common speed after impact will be $\frac{mr+ns}{m+n} \cdot C$ or $\frac{mr-ns}{m+n} \cdot C$, according as to whether they are moving in the same or in opposite directions before Impact (Prop. 3).

In his next proposition Wallis enunciates a theorem in which the idea underlying Newton's Second Law of Motion is even more clearly discernible. The total effect of a force in producing change of momentum is called its Impulse; for this Wallis uses the expression

* *Op. cit.*, Scholium, Prop. 2, Cap. xi.

"*magnitudo Ictus*"—the magnitude of the blow—and Proposition 5 states that the magnitude of the blow is double the momentum lost by the striking body—that is to say, it is equal to the total change of Momentum—the Momentum lost by the one *plus* the momentum gained by the other.

The generalization is given in Proposition 13. A body of mass mP and speed rC impinges upon a body of mass nP moving in the same direction with a speed sC; then the aggregate of the magnitude of the blows is

$$\frac{2mnr - 2mns}{m+n} \cdot \text{PC} = \frac{r-s}{m+n} \cdot 2mn\text{PC}.$$

The magnitude of the blow (which is equal to the total change of momentum) bears to the momentum of either of the bodies, endowed with a velocity equal to the difference between the initial velocities, a ratio equal to twice the mass of the other body to the mass of the two bodies combined (Prop. 12), *e. g.*,

$$\frac{\frac{(r-s)}{m+n} \cdot 2mn\text{PC}}{(r-s)m\text{PC}} = \frac{2n}{m+n},$$

or

$$\frac{\frac{(r-s)}{m+n} \cdot 2mn\text{PC}}{(r-s)n\text{PC}} = \frac{2m}{m+n}.$$

In Proposition 15 he introduces the notion of Centre of Percussion, or as he calls it Centre of Forces, defining it thus: *Centrum Virium, seu Percussionis. Quod ipsum est Punctum Percussionis maximæ*. But here he makes a slip, which every one of his predecessors and contemporaries made, with the sole exception of Huygens, in that he regarded the Centre of Percussion and the Centre of Oscillation as identical. "*Centrum Virium seu Percussionis Quod quidem non aliter differt a Centro Æquilibrii (de quo Cap. 3. dictum est*")*. He was very definite concerning this. He was well aware of what Huygens had declared (although the *Horologium Oscillatorum* did not appear until nearly three years later), for he concludes his chapter with the observation: "Lastly it must be advised that what we call Centre of Forces or Centre of Percussion or even Oscillation is the very same as what Huygens afterwards mentioned as Centre of Oscillation. Indeed it is the same, though under different names with what we both enquire

* "The Centre of Forces or of Percussion, which does not differ from the Centre of Equilibrium, which was spoken of in Chap. 3."

after, he by his method, I mine. His oscillation is the same as our vibration " *.

This conflict of opinions was revived when Huygens' monumental work appeared in 1673. In that year Wallis wrote to Oldenburg: " Huygens .... De Centro Oscillationis, what he brings, is, in effect but just the same with mine put into another dress and spun into length. In my Cap. ii, *De Motu* (which is De *Percussione*), Prop. 15, I had showed the way to calculate the *Centrum Virium* or *Centrum Percussionis* and (that I might not tediously repeat what had been delivered before), after I had showed the method of it in divers examples I say (in the beginning of the *Scholium*) that from the *Centrum Gravitatis* of the figure to collect the Centrum Percussionis was but the same thing as from the *Centrum Gravitatis of a plain* to collect the *Centrum Gravitatis* of its Ungula or Cuneus of which I had discoursed at large in the second part to which I there refer. And in the close of the Scholium: That this *Centrum Percussionis* or Virium, is the same point with Centrum Vibrationis, or (as he calls it) Centrum Oscillationis. And he does so little disguise my method that in his definition 14, 15. Prop. 7 and what follows he expressly hath recourse to my *ungula* and pursues the notion just as I had showed, only he changeth the name, and what I called *Ungula* he calls *Cuneus*, like as what I had called *Centrum Virium seu Percussionis* he calls *Centrum Oscillationis seu Agitationis*" †.

This was typical of Wallis. He seems to have fostered an intense impatience towards criticism, a profound irritability towards those who differed from him. His method of defence was invariably counter attack, and in no case does he appear to have adopted this method with any degree of reluctance. Not often did Wallis admit himself in error. True, he rarely had occasion to do so, but in this instance he did err in regarding the Centre of Percussion and the Centre of Oscillation as synonymous expressions. Huygens ventured to point this out to him, perhaps not very tactfully. Wallis's comment was: "If Mr Hug. be out of humour, I cannot help it" ‡.

In Chapter xiii, *De Elatere et Resilitione seu Reflexione*, Wallis attacks the more formidable problem of Impact of Elastic Bodies.

* "Monendum denique: Id quod nos *Centrum Virium* seu *Centrum Percussionis* aut etiam *Vibrationis* hic appellamus, id ipsum est quod Cl. Huygenius .... appellat *Centrum Oscillationis*. Quippe idem est (utut sub diversis Nominibus) quod uterque inquirimus. Ille quidem sua Methodo, ego mea." *De Motu*, p. 1015. Opera I.
† Wallis to Oldenburg, March 20, 1673/4. (*Royal Society Guard Books*.)
‡ Wallis to Oldenburg, June 22, 1674. (*Royal Society Guard Books*.)

Two definitions are necessary, viz., the force of elasticity and elastic body.

(1) *Vim elasticam appello, eam qua Corpus de figura sua Vi detrusum seipsum in figuram pristinam Restituere satagit.*

(2) *Elaterem appello, Corpus (aut etiam partem Corporis) ea Vi præditum.* That is to say a body is said to be elastic when the force of restitution called into play during compression is sufficient to make it recover its shape after compression. His usual expressions for these terms are "springynesse" and "springy body".

Proposition 1 states : " If a heavy body impinge upon a solid obstacle, and if either one or the other be elastic, the body will rebound with the same speed and in the same right line." This is based upon the assumption that the force exerted by the elastic body in trying to recover its shape is the same as the force which deformed it, so that this force acting upon the striking body will give it its initial speed again. If the impact be oblique Wallis observes that the angle of incidence is equal to the angle of reflection. (Prop. 1.)

Next, if the two striking bodies are unequal, but have their velocities inversely proportional to their masses (so that their momenta are equal), each will rebound with the same speed that it approached, and in the same right line. (Prop. 3.)

If a heavy elastic body impinge directly upon an equally heavy body at rest, the moving body will be brought to rest, and the other will move forward with the velocity possessed by the striking body. (Prop. 5.)

And generally. (Prop. 8.)

If a heavy body moving with any speed whatever impinge upon a heavy body at rest, and if one or the other be elastic, the final speed of the striking body will be to its speed before impact as the difference of the masses to their sum, and will be forward, backward or stationary according as the mass of the striking body is greater than, less than, or equal to that of the body at rest. The stationary body, however, will acquire a speed which bears to the original speed of the striking body the ratio of double the mass of the striking body to the sum of the masses.

All this may be summarized in modern language and notation thus :—When two bodies impinge the sum of their momenta along the line of Impact is the same as before, and (for bodies perfectly elastic) the relative velocity after Impact is the same as that before Impact, though reversed in direction. For if M and m are the masses, U and 0 (zero) their speeds before Impact, V and v their speeds after impact, we have by the principle already enunciated :

$$MV + mv = MU + 0,$$

and
$$V - v = -U,$$

whence
$$V = \frac{MU - mU}{M + m},$$

and therefore
$$\frac{V}{U} = \frac{M - m}{M + m},$$

i. e., $\dfrac{\text{Speed of striking body after Impact}}{\text{Speed before Impact}}$
$$= \frac{\text{Difference of the Masses}}{\text{Sum of the Masses}}.$$

Again, from the above relations
$$v = \frac{MU + MU}{M + m},$$

whence
$$\frac{v}{U} = \frac{2M}{M + m},$$

i. e., $\dfrac{\text{Speed acquired by stationary body}}{\text{Original Speed of striking body}}$
$$= \frac{\text{double the mass of the striking body}}{\text{sum of the masses}}.$$

This was a notable advance upon anything that had gone before, and, save in two respects, was quite equal to what was subsequently delivered by Newton. In the first place, Newton's notation was much more concise, and more readily understandable. Secondly, Newton introduced the coefficient of restitution e, and thus carried Wallis's investigation to its consummation. For, following Wallis, Newton obtained the relation
$$MV + mv = MU + mu,$$
and from this own experimental results
$$V - v = -e(U - u),$$
where M and m are the masses of the bodies, U and u their initial velocities, V and v their final velocities.

An account of Wallis's Treatise on Motion would not be complete without a reference to his investigations upon the Composition of Motions, and upon Accelerated Motions. We pass therefore to the chapter whose title is *De Motibus Compositis Acceleratis, Retardatis, et Projectorum.*

His first proposition gives greater precision to a principle with which Galilei had startled the scientific world, namely, that a constantly applied force produces not simply a velocity but a *change* of velocity, *i. e.*, an acceleration. *Si, mobili in Motu posito, accedat nova Vis, seu novus Impetus, secundum eandem directionem ; sit Motus Acceleratio* *.

If this force act in the opposite direction it will not only bring the body to rest but will give it an acceleration in the reverse direction.   Furthermore (Prop. 2) if this force is uniform (*si Vis Motricis, per se æquabilis, continua fiat applicatio producetur Motus continue Acceleratus*) the acceleration produced will be uniform also, uniformly accelerated motion being defined as that in which equal increments of speed are added in equal increments of time. This he demonstrates graphically, and though his investigation lacks the definiteness of Newton's Second Law, there is in Wallis's work something more than the germ of Newton's.

Now gravity is a constant force ; therefore if we exclude the resistance of the air, a body falling under the action of gravity will be uniformly accelerated, from which it is shown that the distances traversed by such a body are proportional to the duplicate ratio of the times (Prop. 3). This will persist for a body projected in any direction. Not only that, but if a body, instead of falling vertically, roll down an inclined plane, it will acquire the same velocity (and hence the same *Impetus*) as if it had fallen through the same height vertically.

Proposition 6 contains the important principle of the Parallelogram of Velocities—if upon a body free to move there be impressed two velocities which can be represented in magnitude and direction by the angular sides of a parallelogram, the diagonal through that angular point will represent its acquired velocity in magnitude and direction (*feretur Mobile per Parallelogrammi diagonium, ea celeritate quæ sit ad datas, ut diagonium illud ad respectiva latera*).

This is extended in Proposition 7 : If a uniform motion is compounded with an accelerated motion, the motion will deviate from a straight line, and in the case of a body projected at an angle the path described by the body will be a parabola. Of course all this had been enunciated by Galilei ; Wallis's investigation, however, is very much more comprehensive, for, as usual, he generalizes his results, and considers the motion compounded of all kinds of motions, e. g., *si motus æquabilis componatur cum Accelerato in Temporum ratione Duplicata (ut hic sit ad illum in ratione*

---

* " If a body already in motion be acted upon by a new force or a new impetus in the same direction, the motion is accelerated."

Triplicata) *Latio erit in Curva paraboloeidis Cubicalis*; *i. e.*, if a uniform speed be compounded with one which is accelerated in the duplicate ratio of the times, so that the displacement is in the triplicate ratio of the times, the path will be a Cubical Parabola.

Wallis's demonstration that, neglecting the resistance of the air, the path of a projectile is a parabola is as follows : Let the projectile be given a uniform motion along the line A T T T... (fig. 15); it will continue to move (by Prop. 2, Cap. 1) with undiminished speed. Meanwhile, the force of gravity pulls it downwards, giving

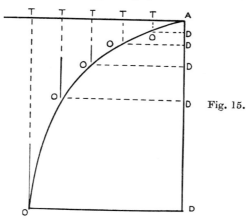

Fig. 15.

it a motion which is uniformly accelerated (by Prop. 3), and this causes it to fall through distances A D D D D... which are proportional to the squares of the times. It is clear that the position of the particle at any instant, under the action of the two motions, can be determined, and it is found to be the curve A O O O O..., which is a parabola. If, however, the air exerts a sensible resistance, as it must do, each of the component motions which is supposed to be uniform (or uniformly accelerated) will not be so. " Otherwise ", says Wallis, " a bullet ought to strike with the same force at the greatest distance as close by the mouth of the piece, which all experience doth contradict. So that I incline to the latter hypothesis which supposeth it compounded of two motions, the one retarded, and the other accelerated .... And practical canoneers, I am told, find the random of a bullet very different from a parabola " *. This was not worked out in *De Motu*, and although Wallis in response to a request from Halley promised to investigate the problem and to communicate his results to the

* Wallis to Collins, Oxford, Aug. 24, 1674. (*Royal Society Guard Books*.)

Society, he was forestalled by Newton. " By those papers of Mr. Newton ", he wrote to Halley, " I find he hath considered the Measure of the Air's resistance to bodies moved in it, Which is the thing I suggested in one of my late letters. . . . I should have proceeded on the same principle: That the resistance (cæteris paribus) is proportional to the celerity (Because in such proportion is the quantity of air to be removed in equal times) " *.

The publication of this monumental work revived interest in mechanics. From being the most neglected of the sciences, mechanics rapidly became not only one of the most powerful, but also one of the most intensely cultivated. This sudden accession to popularity was, of course, in a large measure due to Newton, but there can be no doubt that just as Newton owed much in his mathematical researches to Wallis, his obligations in mechanics were not less pronounced.

* Wallis to Halley, December 14, 1686. (*Royal Society Guard Books.*)

CHAPTER VIII.

The Hooke-Hevelius Dispute.
Publication of Ancient Manuscripts.

THE controversies which blazed around the scholars of the seventeenth century usually had far-reaching effects, inasmuch as the disputants did not for long lack partisans. No better illustration of this melancholy state of things can perhaps be cited than the quarrel which arose between Hooke and Hevelius concerning the respective advantages of plain and telescopic sights. This quarrel had not advanced very far before other distinguished scholars, notably Wallis, ranged themselves on one side or other of the contending parties.

The dispute seems to have been occasioned by the publication of Hevelius's *Cometographia* (Gedani, 1668), a copy of which had been sent to Hooke as well as to other members of the Royal Society. Hooke in return sent Hevelius a description of his Dioptric Telescope, and he called attention to the superiority of the observations made with such an instrument over those made with plain sights. But Hevelius, perhaps because he felt himself too old to change, continued to rely upon plain sights. This led to not a little unpleasantness between the two astronomers, the point at issue being whether distances and altitudes could be taken with plain sights nearer than to a minute. Hooke, the champion of telescopic sights, even went so far as to suggest that "it were not possible, with these *Sights*, (be the Instruments never so large or accurate,) to make Observations nearer then to Two or Three whole Minutes : But himself could, with *Telescopick Sights* ; (by an Instrument but of a Span breadth,) make observations, Thirty, Forty, Fifty, yea Sixty times more accurate than could be done the other way, with the most Vast Instruments " *.

As might be expected, the dispute was carried on with considerable vehemence. Hevelius, by way of a challenge, published his *Machina Cœlestis, Pars Prior* (Dantzic, 1673), in which he gave a description of his instruments, together with a series of observations stretching over half a century. Copies of this were sent to different members of the Royal Society (November, 1673), and many of them, notably Halley, expressed great admiration for the work. Hooke, however, received no copy. At once he challenged Hevelius's work.

* *Phil. Trans.*, 1685, pp. 1164/5.

With a lack of restraint not unusual in him he devoted several Cutlerian lectures to hostile comments on that " curious and pompous Book of the first part of his *Machina Cœlestis* " *. The next year he gathered together his criticisms and published them in his *Animadversions on the First Part of the Machina Cœlestis* (1674), a treatise replete with unhandsome reflections upon the Dantzic astronomer. Hevelius was highly incensed at the pontifical manner adopted by Hooke, and in a letter to Oldenburg, which was read to the Society on April 23, 1674, he accused his adversary of " much more of bitterness and boasting than there was reason for. Which he thinks was done out of design to disparage Him, his Instruments and his Observations .... thus seeking to raise his own reputation by disparaging what is done by others in things wherein himself doth nothing " †. Wallis, who from his letters had joined in the dispute many years earlier, now wrote to Oldenburg : " I have now read the whole of Mr. Hooke's against Hevelius, which I think bears a little too hard upon him. For Hevelius hath deserved well " ‡, and a subsequent letter showed that he still rated the achievements of the Dantzic astronomer very highly. " We have no reason to be displeased with him ", he wrote, " We are to consider that his Instruments were made with great cost and care, and he a diligent observer with them, and by long practice expert in the manage of them, long before these telescopick sights were thought of " §. In 1679 Halley accepted an invitation to visit Hevelius with a view to arbitration in the dispute, and in a letter which he left behind he declared himself " abundantly satisfied with the use and certainty of these Instruments, and offers himself a voluntary witness (of the almost incredible certainty of his Instruments) against all who shall for the future call his Observations in question " ‖. Hevelius's results were eventually published in his *Annus Climactericus* (Gedani, 1685), and again they met with a hostile reception from Hooke. Wallis, however, on being consulted gave a highly favourable review, in which he maintained that Hevelius had been misrepresented not only by Hooke but also by Flamsteed. " As to Mr. Halley ", he wrote, " if you think, (as you seem to intimate) that he hath been too lavish in his commendations, you must think Mr. Hook hath been so in his reprehensions. Nor are the instruments and observations so contemptible even in our judgment as he seems to represent them .... But I think this is undeniably evinced that

* *The Posthumous Works of Robert Hooke*, Rd. Waller, Lond., 1705, p. xv.
  † *Phil. Trans.*, 1685, p. 1165.
  ‡ Wallis to Oldenburg, Jan. 11, 1673/4. (*Royal Society Guard Books*.)
  § Wallis to Oldenburg, Jan. 12, 1673/4. (*Royal Society Guard Books*.)
  ‖ Halley to Hevelius, July 8/18, 1679. *Phil. Trans.*, 1685, p. 1169.

Hevelius with his plain sights can distinguish to a small part of a minute, notwithstanding what hath been said to the contrary " *.

No one could ever maintain, however, that Wallis's knowledge of astronomy was reliable. He quite believed, for example, that Flamsteed had observed a parallax of the Pole Star, and not only did he allow this to be published in his *Opera*, he actually called attention to it in a letter to Pepys. " You will find in the collection of letters ", he wrote, " some remarkable observations of Parallax of the earth's annual orb by Mr. Flamsteed (in confirmation of the Copernican Hypothesis) a thing long sought for, and never observed till now " †. But he was acute enough to realize the worth of the works of young Jeremiah Horrocks. " The Latin pieces ", he wrote to Oldenburg, " we, (Doctor Wren and myself) look upon as the beginning and attempt of an excellent work for the restitution of Astronomy, and which serves to show how great a losse it was that he died so soon, who gave evidence by this . . . . with how much diligence and sagacity he was like to have applyed himself to these studies had he lived " ‡. So impressed was he, in fact, with those works, that he himself saw them through the press, and they eventually appeared in 1673 under the title *Horrocii Opera Posthuma*.

Meanwhile Wallis had already embarked upon the task of correcting and emending for his own private use the manuscript copies of *Archimedis: Arenarius, et Dimensio Circuli*. This treatise had already been printed twice in Greek and thrice in Latin, but as there were many defects in the copies then extant Wallis supplied another edition to which he added copious explanatory notes. This first appeared in 1675, and though it had been intended primarily for his own use, it lacked not the minute care and precision which had characterized all his other mathematical writings. Although Wallis had done more than any other writer to displace the old synthetic methods of the ancients, and although he had again and again severely criticized the mathematicians of antiquity, whom he likened to " Builders, who, when the house is finished, take away the Scaffolds imployed in the work " §, nevertheless he seems to have had a profound admiration for their writings. For when Collins suggested his bringing out an abridgment of the mechanical treatises of Aristotle, he replied : " I had much rather hearken to ye printing those Antients at large in Greek and Latine wch hath yet never been done, and ye Authors thereby in danger of being lost "‖. Although the project did not meet with much

* Wallis to Oldenburg, Feb. 12, 1685/6. (*B. B.* 2498.)
† Add. MS. no. 113 (Bodleian).
‡ Wallis to Oldenburg, April 6, 1664. (*Royal Society Guard Books*.)
§ *Algebra*, p. 213.
‖ Wallis to Collins, Aug. 24, 1674. (*Royal Society Guard Books*.)

encouragement, he published in 1680, from the MSS. *Claudii Ptolemæi Harmonicorum Libri Tres* in Greek with a Latin version and notes, to which he afterwards added an Appendix : *De Veterum Harmonica ad Hodiernam Comparata* as well as *Porphyrii in Harmonica Ptolemæi Commentarius*. Wallis endeavoured to account for the surprising effects attributed to the ancient music, e. g., Timotheus exciting Alexander's fury with the Phrygian mode, or soothing him to indolence with the Lydian, and he ascribes them chiefly to the novelty of the art and to the exaggeration of the ancient writers. Nor does he doubt but that modern music could produce effects as considerable. In this work Wallis rendered a useful service by reducing the ancient music to modern notation. Ptolemy's treatise gave an account of the nature of sounds in general, especially such as were musical, and he discoursed on the relations existing between the notes of the musical scale. He discussed the imperfections of the Organ, due to the fact that since all the semitones were equal in value, it was impossible for these to coincide with the notes of the natural scale. This formed the subject of two letters which were published in the *Transactions* for 1698. Yet in spite of this, and many other writings on the theory of music, he knew nothing of the practice *. In July 1688 he published *Aristarchi Samii* : *De Magnitudinibus et Distantiis Solis et Lunæ*, to which he subjoined *Pappi Alexandrini, Libri Secundi Collectionum Mathematicarum (hactenus desiderati) Fragmentum*. These form the bulk of the third volume of his *Opera* (1699), concerning which the *Philosophical Transactions* contains the following eulogy : " Much of the present volume is employed in preserving and restoring several ancient Greek authors who were in danger of being lost. For which work the Doctor is fitted, not only by his excellent knowledge in Mathematics, accurateness in the languages and great industry in collating manuscript copies, but also by what is peculiar to him, his art and practice of decyphering, which enables him to make sagacious conjectures, supplements and emendations, which must often be an editor's business, and which we so fully admire in him " †.

In 1685 he published his *Algebra* in English This was reprinted with copious additions in the second volume of his works which were printed between 1693 and 1699. A large portion of the treatise is taken up with the history of the subject, and it shows an extraordinarily intimate acquaintance with the ancient mathematical works. This treatise also contains an account of the work which had been done subsequent to his own *Arithmetica Infinitorum*.

* Hearne, pp. 67, 73.
† *Phil. Trans.* (abridged), Hutton, vol. iv, p. 410.

Wallis called particular attention to Newton's brilliant discovery of the Method of Fluxions. The Differential Calculus of Leibniz was not even mentioned, and for this omission Wallis excused himself by stating that it was but the same as Newton's Method of Fluxions *. An account of the *Algebra* is given elsewhere.

In 1687 he published his *Institutio Logicæ* (Oxon, 1687). " This Institution of Logic ", wrote Doctor Gregory, " came very seasonably from so great a master in Mathematical and Physical learning, as he was as a rebuke to those who under colour and pretence of advancing those usefull sciences were running down the Logick and Metaphysick of the Schools " †. Wallis was much more emphatic regarding the value of the treatise he was about to publish, which, he says, " differs from other treatises because my design is something different from theirs, being to obviate a mistake which I find some apt to run into. As if the whole business of Logick were but to dispute or quarrel about Predicamental or Transcendental with other great words which they learn and which were of no further use in human life. So it is the business of Logick .... to manage our Reason to the best advantage, with strength of Argument and in good Order, and to apprehend distinctly the strength or weakness of another's Discourse, and discover Fallacies .... To show that what some apprehend mostly as an Art of wrangling or useless canting, is a thing of universal use in all rational discourse " ‡.

In 1690 he published a treatise *The Doctrine of the Ever Blessed Trinity Explained*, which involved him in a dispute with the Unitarians, and as soon as this had run its course he engaged in a controversy with a Thomas Bampfield in defence of the Christian Sabbath.

The proposal for reforming the Calendar was introduced about this time. The Gregorian Calendar, it may be recalled, had been adopted in Italy, France, Spain, Portugal and Poland in 1582. Its adoption throughout the rest of Europe was gradual, but in 1700 it was in use in practically every European country except our own. When in that year the German and Dutch Protestant States and Denmark decided upon its adoption, a move was made to bring this country into line with the rest of Europe, and Wallis, in virtue of the position he had now attained, was consulted. He strenuously opposed the suggested alteration, alleging that not only would it give rise to great confusion in astronomical calculations, but would unnecessarily disturb the ordinary natural transactions.

* *De Morgan*, p. 26.
† Gregory, *Life of Wallis*. Smith MS. 31 (Bodleian).
‡ Wallis's Letter, read to the Society, May 30, 1685.

Moreover, as he pointed out in a letter to Sloane *, the Gregorian or New Style did not exactly conform to the motions of the celestial bodies, and consequently would itself, in process of time, stand in need of correction. But his main objection was that the design had been sponsored by a Roman Pontiff. "It had been much better", he wrote in the *Transactions*, "if it (the old style) had so continued to this day, rather than Pope Gregory on his own single authority should take upon him to impose a law on all the Churches and Kingdoms and States of Christendom to alter both the ecclesiastical and civil year for a worse form than what we had before" †. Moreover in a letter to Doctor William Lloyd, Bishop of Worcester, he expressed strong disapproval of the proposed change ‡, the adoption of which would greatly encourage the Roman Catholic party, who would not, he alleged, fail to make use of this as an argument that we were coming over to Rome ! Wallis's counsels prevailed, and the fact that such a useful design was for a time further laid aside, and indeed not adopted until 1752, speaks eloquently of the reputation he had now acquired amongst his own contemporaries.

* Wallis to Sloane, May 11, 1700. (*Royal Society Guard Books*).
† *Phil. Trans.* (abridged), Hutton, vol. iv, pp. 434/5.
‡ *Tanner Papers*, xxi, June 30, 1699 (Bodleian).

CHAPTER IX.

THE *TREATISE OF ALGEBRA*.

THOUGH he was approaching his seventieth year, Wallis now turned aside from these activities and began to direct his energies to the production of his erudite *Treatise of Algebra*. This important work, which was written in English, appeared in 1685, and of all the author's vast output it was probably during the next hundred years or more the most widely read. It rapidly became a standard text-book on the subject, and, largely on account of the improved notation which Wallis adopted in its pages, it soon displaced many of the treatises then current.

It is, however, not merely as a treatise on algebra that the work claims attention ; the book marks the beginning, in England at least, of the serious study of the history of mathematics. " It contains ", says Wallis in his Preface, " an Account of the Original, Progress and Advancement of (what we now call) *Algebra*, from time to time, showing its true Antiquity (as far as I have been able to trace it ;) and by what steps it hath attained to the Height at which now it is " *. His account of the history of mathematics in antiquity is very comprehensive and gives evidence of a close study of the classical literature of the science. But when he approaches his own day, he displays a marked tendency to overrate the achievements of his fellow-countrymen, and in consequence his account has been provocative of many conflicting opinions. Rouse Ball, for example, claims that Wallis displays scrupulous impartiality in attributing the credit of different discoveries to their true sources †, an opinion which is shared by not a few English writers. Montucla, on the other hand, strongly protests against its unfairness to the early French algebraists, particularly Vieta and Descartes. " Il fut toujours peu favorable, pour ne rien dire plus, aux François et à Descartes en particulier ", asserts that distinguished historian ‡. De Morgan, while admitting considerable merit to the work, shares Montucla's opinion, and Cajori's estimate of the historical treatment is that it is unreliable and worthless. A most cursory examination of the work reveals quite clearly that the criticism of these latter is well grounded,

* Preface to *Treatise of Algebra*, London, 1685.
† Ball (Camb.), p. 45.
‡ Montucla, ii, p. 349.

and that Wallis, though he claimed "to have endeavoured all along to be just to everyone", failed lamentably in his attempt.

It can scarcely be doubted that Descartes' *La Géométrie* (1637) was an epoch-making work, and one which, even apart from its exposition of analytical geometry, did a great deal to extend the bounds of algebraical knowledge. De Morgan* declares that it was this work which inspired Newton with an affection for the subject, and Cantor describes it as a milestone in the development of the theory of equations ("Einen wirklichen Markstein in der geschichtlichen Entwickelung der Lehre von den Gleichungen") †. Descartes was an able mathematician, and no greater testimony to his work can be found than that it has survived the vicissitudes of nearly three centuries. Yet although Wallis takes pains to mention the improvements made by Leonardo of Pisa, Tartaglia, Paccioli, Bombelli, Vieta and many others, he persistently refrains from mentioning Descartes, except to compare him, and that unfavourably, with Harriot. "Harriot .... in sum, he hath taught (in a manner) all that which hath since passed for the *Cartesian* method of Algebra; there being scarce anything of (pure) *Algebra* in *Des Cartes* which was not before in *Harriot*; from whom *Des Cartes* seems to have taken what he hath, (that is purely *Algebra*), but without naming him. But the application thereof to *Geometry* or other particular subjects (which *Des Cartes* pursues) is not the business of that treatise of *Harriot* (but what he hath handled in other Writings of his, which have not yet had the good hap to be made publick), the design of this being purely *Algebra*, abstract for particular Subjects. Of this Treatise here is a fuller account inserted because the Book it self hath been but little known abroad, that it may hence appear to what estate *Harriot* had brought *Algebra* before his death" ‡. And although he maintains that he has tried "all along to represent the sentiments of others with Candour, and to the best advantage; not studiously seeking opportunities of Cavilling, or greedily catching at them if offered" §, he actually declares that Harriot "hath laid the foundation on which *Des Cartes* (though without naming him) hath built the greatest part (if not the whole) of his *Algebra* or *Geometry*. Without which, that whole Superstructure of *Des Cartes* (I doubt) had never been" ||.

Not unnaturally, such an extravagant claim raised a sharp controversy between the English and the French mathematicians,

* De Morgan, p. 11.
† Cantor, ii, p. 722.
‡ Preface to the *Algebra*.
§ *Ibid.*
|| *Algebra*, p. 126.

CHAPTER IX

nor was this controversy confined to Wallis's own age. "Comment excuserons-nous M. Wallis", is the indignant complaint of Montucla, " qui nous donnant un Traité historique de l'Algèbre, semble avoir à peine jetté les yeux sur tout autre analyste que Harriot ; qui après avoir traité Descartes de plagiaire, et avoir déprimé ses inventions autant qu'il l'a pu, forme en grande partie l'énumération de celles de son compatriote, de choses ou peu importantes, ou empruntées de ses prédécesseurs. Qui pourra même ne pas rire en voyant ce zélé restaurateur de la gloire d'Harriot, lui attribuer, je ne dis pas seulement la résolution des équations du second degré, par l'évanouissement du second terme, invention de Viète, mais encore la méthode vulgaire qui procède, comme on sait, en ajoutant de part et d'autre de quoi faire un quarré parfait du membre où est l'inconnue. La partialité et l'aveuglement qui en est la suite ordinaire ne sauroient être portés plus loin " *.

This is a bitter indictment of Wallis's claim to have ascribed " every step of Advance to its own Author ". But it is not unjust. For according to Wallis, Harriot was responsible for so many " strange improvements " and was so little taken notice of by foreigners that it had now become a duty to describe more fully the achievements of which Harriot was the author but which now passed under other names !

It is not easy to understand why one whose knowledge of the history of mathematics was so profound, and whose reputation was so firmly established, should have been guilty of such gross partiality. Montucla suggests that Wallis's antagonism towards Descartes was the outcome of his disputes with Pascal, Fermat and other members of the French mathematical school. Certainly his hostility towards Descartes seems to have developed only in later years ; in the *Arithmetica Infinitorum* (1656), for example (Scholium Prop. 107), he classes him with the greatest mathematicians of the age. But whatever truth there is in Montucla's suggestion, there can be no doubt that Wallis's praise for Harriot, whom he likens to Columbus the discoverer of a New World, is monstrously overstated, whilst his strictures on Descartes are unreasonably severe. In the Preface to the *Algebra* in vol. ii (1693) of the *Opera*, Wallis, speaking of Descartes, wrote *cui non sum inimicus ego* ; if that be so, his censures would seem to have the saving grace of sincerity. Unfortunately, this attitude finds no support amongst his other statements. Among the Rigaud MSS. in the Bodleian are to be found the following *Extracts from some Memoranda in a Blank Leaf of Harriot's Artis Analyticæ Praxis*: " This Treatise of Mr. Harriot ", wrote Wallis, " was (it seems) so well liked by Des Chartes that he hath

* Montucla, ii, pp. 110/111.

in a manner transcribed the whole of it for the substance (though in other order and words) into his *Geometry* (but without so much as even naming the author), which was first published in the year 1637 (in French) six years after this was first extant. There are many other worthy pieces of Mr. Harriot's doing left behind and well worth publishing "*. It will be shown later that for this last statement Wallis had no evidence whatever; in the pages of the *Algebra*, however, he tries to justify the claims he has made, and although he shows himself a zealous champion, it cannot be said that his claims will bear even the slightest scrutiny.

Thus, maintains Wallis, in his treatment of equations, Harriot takes into consideration negative, or privative, roots which by some are neglected. In all this he is followed by Descartes, save that what Harriot very properly called privative roots, Descartes is pleased to call false roots. When, eleven years earlier, Wallis had written to Collins concerning Descartes' *Geometry*, he had accused its author of having taken the idea of negative roots from him, and not from Harriot : " I found also my negative roots owned by him under the name of false roots, though they are indeed as true as the other " †. Wallis seems to suggest that, because Descartes described these roots as false, he was ignorant of their nature. But the frequent use which Descartes makes of them in *La Géométrie* ought to have dispelled that notion from Wallis's mind. Was not Descartes, to say the least, familiar with the theorem by which the number of positive and negative roots in an equation are determined ? Indeed, in *La Géométrie* there are not wanting instances of Descartes' ability to handle imaginary roots ; that he should be unfamiliar with the nature of negative roots we find extremely difficult to believe. In fact, if Montucla is right, it is to Descartes that we are indebted for a knowledge of such roots, since he was the first to make extensive use of them : " C'est à Descartes, nous le répéterons ici, qu'est due la connoissance de la nature et de l'usage des racines négatives, et il est le premier qui les ait introduites dans la géométrie et dans l'analyse " ‡. Moreover, this view is supported by our own countryman Thomson, in his erudite *History of the Royal Society*, where he says : " It was he [Descartes] that first pointed out the importance of the negative roots of equations ; and he added an important fact to the theory of Harriot, respecting equations " §.

In all his work on equations, continues Wallis, Descartes follows Harriot, or rather borrows from him. Harriot's work on equations,

* Rigaud MSS. 20 (Bodleian), vi, p. 41.
† Wallis to Collins, April 12, 1673 (Rigaud, vol. ii, p. 573).
‡ Montucla, ii, p. 114.
§ Thomson, p. 270.

wherein he shows the formation of higher equations from a composition of lateral or more simple equations (which, says Wallis, is the great key that opens the most abstruse mysteries of algebra, and which we owe purely to him), was so highly appreciated by Descartes that in his *Géométrie* he made no scruple about following him almost in everything, adding very little of his own. In fine, Descartes gives only one rule which Wallis cannot trace back to Harriot, namely, the " rule for resolving a biquadratic Equation whose second term is wanting into two Quadratics by the help of a Cubic Equation of a Plain Root ". This, maintains Wallis, is the only thing which he adds to what we have related out of Harriot. But even in this Wallis will not accord to Descartes the credit of originality, for he hastens to add that Descartes' rule " differs not in Substance from those of Bombel and Vieta " *. To what extent we can rely on Wallis's judgment becomes only too plain when we refer to his letters. In 1673 he wrote to Collins : " In the year 1648, in answer to a letter of Mr. Smith, which was the first occasion of my sight of Descartes' *Géométrie*, then extant only in French, being by him desired to give him an account of it, I (because he saith nothing of the way whereby he came at it) set myself to find a rule to do it [*i. e.*, to reduce a biquadratic equation to a cubic ; Descartes' *Géométrie*, Liber 3, p. 79], which proved to be the same with his, and in demonstrating it, I did his also, which I have since communicated to Doctor Twysden and others " †.

How far Descartes merited Wallis's strictures and what foundation there was for his oft repeated charges of plagiarism can be judged only by a closer study of the " strange improvements " which Wallis so frequently asserts were Harriot's legacy to algebra. Of Wallis's earnestness in trying to establish the claims on behalf of his fellow-countryman there can be no doubt whatever, for more than a third of the treatise is devoted to an account of the achievements of Harriot, an account so eulogistic that he almost passes from the language of panegyric to that of homage.

It was in notation that Harriot rendered notable service to the development of algebra. The symbols $>$ and $<$ are due to him. The first improvement, however, to which Wallis directs attention, is his use of small letters, as taking up less room. Harriot abandoned the terms square, cube, sursolid, etc., and he substituted a, aa, aaa, $aaaa$, for what had hitherto passed as A, Aq, Ac, Aqq, " . . . . which he performs more naturally by the bare Number of their dimensions, as a, aa, aaa, ... which, when they come to be numerous, it is

* *Algebra*, p. 208.
† Wallis to Collins, Mar. 29, 1673 (Rigaud, ii, p. 559).

conveniently expressed by a numeral figure adjoyned, as a^3, a^4, instead of *aaa, aaaa*, ... which Mr. Oughtred did sometimes also use " *. If by this Wallis means to suggest that Harriot used the index notation he is certainly in error, for there is not a single example of it in the pages of the *Praxis*. Moreover, the nearest that Oughtred ever came to it is in the *Clavis*, Cap. xii, where he says: " A cube multiplied into its Length makes a Square-square, whose Power is 4, this again multiplied into a Side similarly makes a Quadrato-Cube, whose power is 5 " †. Descartes is not mentioned, which shows how thoroughly he had fallen from grace since the publication of the *Mathesis Universalis*, nearly thirty years earlier, for in the early pages of that treatise Wallis had written: " Descartes and others after him, so as to avoid tiresome repetition of letters, indicated the root by a certain letter of the alphabet as formerly, and the power of it by raised numerical figures (for the number of the degree or power) as a, a^2, a^3, etc." ‡.

It was, however, his work on equations that inspired Wallis to hail Harriot as a pioneer. " He hath also made a strange improvement of Algebra ", he says in his Preface, " by discovering the true construction of *Compound Equations*, and how they may be raised by multiplication of Simple Equations, and may be therefore resolved into such ". Harriot introduced the important principle that every compound equation results from the continued multiplication of as many simple ones as there are units in the index of the highest power, and that consequently the equation has as many roots as it has dimensions. His device consisted in starting with a simple equation, and, by transposing all the terms to one side, equating them to zero. Thus he would write the simple equation $a=b$ as $a-b=0$. Now if this is multiplied by a similar equation, $a+c=0$, we arrive at the equation:

$$aa-ba+ca-bc=0.$$

This is called an *Original Equation*, and the equation which arises by transposing the absolute term to the other side, *e. g.*,

$$aa-ba+ca=bc,$$

* *Algebra*, p. 199.

† " Cubus ductus in latus suum facit quadrato-quadratum, quæ potestas est quartana [4] : hæc iterum ducta in latus facit quadrato-cubum, scilicet quintanam [5] etc." (Oughtred, *Clavis Mathematicæ*, Oxford, 3rd edn., 1652, Cap. xii, p. 34).

‡ " D. des Cartes et post illum alii, literarum sæpe iterandarum tædium timentes, radicem, ut prius qualibet Alphabeti litera designant, et ipsius reliquas potestates suspensis notis numericis (pro numero Gradus seu Potestatis) designant, ut a, a^2, a^3, etc." (*Mathesis Universalis*, Oxford, 1657, p. 71).

CHAPTER IX

is called a *Primary Canonical Equation* (Def. 15, Art. Anal. Prax., 1631). From this latter is derived the *Secondary Canonical Equation*, by expelling the second term, as

$$aa = bb.$$

Similarly, from the three " simples "

$$a = b,$$
$$a = c,$$
$$a = -d,$$

he derives (Sect. iii, Pr. 2, p. 29) the Primary Canonical

$$aaa - baa + bca$$
$$-caa - bda$$
$$+daa - cda = -bcd,$$

from which is derived the Secondary Canonical

$$aaa - bba$$
$$-bca$$
$$-cca = -bbc$$
$$-bcc.$$

Now, by comparison with these canonical equations, which serve as patterns, any given equation, or, as he calls it, any *common* equation, may be solved. Continuing his process of building up the " simples", Harriot proceeds to equations of higher degrees, and thus he is able to demonstrate many of the fundamental properties of these equations, viz. :—

(1) The number of roots of an equation is equal to the dimension of the highest term. This he shows by comparing the *common* equation with the like form amongst his canonicals, which two equations, having the same number of terms and relations between the coefficients, he calls *equipollents*. "These two equations [*i. e.*, the two Canonical Forms] are regarded as Patterns, since by comparison with them may be determined the number of roots in a given Common Equation " *. But for some inscrutable reason he confined himself to those equations that have positive roots only, and should an equation be devoid of positive roots, then, for Harriot, such an equation is *impossible*.

(2) The absolute term (of the Canonical Equation) is the product of the roots.

* " Hæ duæ æquationum species pro canonicis habentur, quia per earum applicationem tanquam per canones sive regulas, radicum numerus in æquationibus communibus determinatur " (Harriot, Def. 17, p. 5).

(3) Not only is it manifest from these compositions how many roots every equation contains, it is likewise clear how the coefficients of the different powers are made up. Thus, having equated all the terms to zero, it follows, as Wallis easily shows, that the coefficient of the second term is the aggregate of all the roots with the sign changed; the coefficient of the third term is the aggregate of all the rectangles made by the multiplication of every two such roots in every possible combination, and so on. By the application of this rule, it is easy to see what are the conditions that any term shall be wanting; conversely, if a particular term be missing, we learn something as to the nature of the roots. All this, maintains Wallis, is evident upon the first inspection of such compositions and is a great advancement towards the perfect understanding of the true nature of each equation.

(4) In his Sixth Section are two other " improvements ", which, says Wallis, we owe entirely to Harriot, firstly, his device of multiplying or dividing the roots of an equation, whilst yet unknown, in any proportion at pleasure, thus freeing the coefficient of an equation of fractions or surds, and secondly, his method of removing one (or more) terms of an equation, thereby reducing it to a smaller number of terms (*e. g.*, an affected quadratic to a simple quadratic).

In all, Wallis gives no less than five and twenty " Improvements of Algebra ", and he actually affirms that these are all explicitly delivered by Harriot either in express words or by obvious remarks upon the bare inspection of what he delivers. " And most of them are properly his own discoveries (for ought I can yet find), though in some few of them *Vieta* had gone before him " *.

This is an imposing list, and one which would undeniably entitle Harriot to be numbered among the greatest mathematicians of even that illustrious age. But there can be little doubt that Wallis has magnified out of all proportion the achievements of his fellow-countryman. Harriot's observation that equations of any order are merely the product of simple equations and hence that his " common " equations could be solved by comparison with his " canonical " forms constitutes a discovery worthy of the ablest analysts of the day. Nor does it detract from the importance of the discovery to say, as does Montucla, that the ground had been so well prepared by Cardan, Vieta, and Descartes that it could not have escaped discovery much longer. One can therefore understand Wallis's waxing enthusiastic over the achievements of his fellow-countryman. But Wallis's other claims on behalf of Harriot will not bear scrutiny. The relations between the

* *Algebra*, p. 200.

coefficients of an equation and its roots, for example, appear in writers before Harriot. Peletarius had observed as early as 1588 that the root of an equation was always a divisor of the last term. Still less can we attribute to Harriot the method of removing one (or more) terms from an equation. But indisputably the greatest blemish on Harriot's treatise is the persistence with which he ignores those equations that give negative roots only. " He finds ", says Wallis, " every Quadratick Equation to have two (Real) Roots . . . . that is, Two Affirmatives, or Two Negatives, or one Affirmative and the other Negative " *. But Harriot actually does nothing of the kind. What he does say (Sect. iv, Pr. 1) is: "Æquationis $aa-ba+ca=bc$, est b radix radici quæsititiæ a æqualis ", $i.\,e.$, in the above-mentioned equation (in which a is the unknown), the root is $a=b$. Not only is there no mention of the other root, but it is actually demonstrated in the Lemma that there is no other root! " Quod autem non detur radix alia præter b æquationis radici a æqualis in sequenti Lemmate demonstratur." And although Harriot deals with a large number of equations in this and in subsequent sections, they are all such that the roots are all positive, or, if any of the roots are negative, they are ignored. For example, in Section v, $i.\,e.$, the section " in which is determined the number of roots of Common Equations by comparison with their Canonicals " †, there occur the following propositions :

(1) " Æquatio communis $aaa-3\,.\,bba=+2\,.\,ccc$ in qua $c>b$ de simplici radice explicabilis est " (Sect. Quinta, Prop. 1), $i.\,e.$, an equation $a^3-3b^2a=2c^3$ is satisfied by one root if b is less than c ; as a matter of fact, in that case there is only one positive root. This he compares with its canonical, $a^3-3rqa=r^3+q^3$, whose root has been shown (Sect. iv, Prop. 14) to be $a=q+r$.

(2) " Æquatio communis $aaa-3\,.\,bba=+2\,.\,ccc$, in qua $c=b$ de simplici radice explicabilis est " (ibid., Prop. 3), $i.\,e.$, the equation

$$a^3-3b^2a=2c^3$$

is satisfied by one root if $c=b$. For its canonical equation is

$$a^3-3q^2a=2q^3,$$

whose root has already been shown (Sect. iv, Pr. 17) to be $a=2q$.

(3) " Æquatio communis $aaa-3\,.\,bba=-2\,.\,ccc$ in qua $b>c$ de duplici radice explicabilis est " (ibid., Prop. 4), $i.\,e.$, the equation

$$a^3-3b^2a=-2c^3$$

is satisfied by two roots when c is less than b, whereas, of course,

* Algebra, p. 131.

† Ibid., Sectio v. In qua æquationum communium per canonicarum æquipollentiam, radicum numerus determinatur.

in that particular case the equation has two positive roots and one negative. The canonical equation with which Harriot compares this equation is

$$a^3 - aq^2 - aqr - ar^2 = -q^2r - qr^2,$$

whose roots have been shown to be $a=q$ and $a=r$.

It is not at all clear why Harriot should so persistently ignore negative roots. No wonder his work on equations failed to impress either Montucla or Cantor! Montucla observes: " J'étonnerai sans doute plusieurs de mes lecteurs lorsque je remarquerai encore qu'Harriot n'eut qu'une idée peu développée des racines négatives " *. Cantor goes much further, for he asserts that positive roots alone have any meaning for Harriot, who actually demonstrates that equations have only positive roots! " Von negativen Gleichungswurzeln will Harriot nichts wissen, nur positive haben für ihn einen Sinn. Ja, er beweist sogar, dass Gleichungen nur positive Wurzeln besitzen ! " †.

Nevertheless, Harriot had some knowledge of the existence of negative roots. His investigation into the formation of equations had taught him how many roots the compound equation should have. If his " solution " did not provide him with all these, he knew how many were missing. How then did he reconcile his rule for determining the number of roots which the equation possessed with the number which emerged ? The answer is in Sect. ii, Prop. 15, where he speaks of such roots as privative ‡, but he hastily passes them by, since, as he says, they are useless and can therefore be neglected. And so he ignores them. But the fact that he should reject as useless equations which are composed of these privatives is no slight deficiency, for the nature of these roots had been investigated by Cardan, whose *Ars Magna* (Nuremberg, 1545) had appeared nearly ninety years earlier. Even imaginary roots seem to have been understood by not a few writers of the period ; Napier, for example, appears from the posthumous *de Arte Logistica* (Edinburgh, 1839, pp. 86-7) to have been cognisant of such, for he draws a clear distinction between the roots of positive and of negative numbers, and he expressly warns against the error of supposing $\sqcup -9$ to be the same as $-\sqcup 9$. All this supports what

* Montucla, ii, p. 107.
† Cantor, ii, p. 792.
‡ " Canonicarum vero derivationes ab his originalibus .... cum absque radicum privativarum suppositione fieri nequeant, tanquam inutiles negliguntur ", *i. e.*, the derivatives of these canonicals from their originals, when the operation cannot be performed without the supposition of privative roots, are, however, neglected as useless (*op. cit.*, Sectio ii, Prop. 15, p. 27).

CHAPTER IX

we have persistently maintained, namely, that Harriot's mathematical ability rarely rose above mediocrity, and that Wallis, in finding in his work what certainly was not there, made a blunder which is difficult to explain and impossible to overlook. Wallis was not unconscious of this blemish in Harriot's work and he tried to excuse it, and perhaps himself also. " In his fourth Section ", he wrote, " Speaking particularly of Affirmative Roots, he doth severally demonstrate .... that such and so many Roots there are (as is above declared) and no more. And having shewed it as to the Affirmative Roots, it may by like Methods, be shewed as to the Negative also " *.

But Wallis's extraordinary claims did not end even here. Not content with ascribing to Harriot a knowledge of negative roots, which was a gross exaggeration of what that writer actually possessed, he now makes equally startling claims concerning his knowledge of imaginary roots. In Chapter xxxii, in discussing imaginary roots and their possible uses, Wallis declares : " And of such imaginary Roots we find Mr. *Harriot* particularly to take notice (in the solution of Cubick Equations) in his 13th example of his Sixth Section ".

Let us see to what extent Harriot " doth particularly take notice " of imaginary roots. The 13th problem of the Sixth Section is :

" Æquationem $aaa - 3.bba = +2 \cdot ccc$, posito $a = \dfrac{ee+bb}{e}$ si c. major sit quam b, ad æquationem simplicem $eee = ccc + \ldots ddd$; Si $c=b$ ad æquationem item simplicem $eee = ccc$. Si vero c minor sit quam b, ad æquationem $eee = ccc + \sqrt{-dddddd}$ impossibilem reducere." That is, by substituting $a = \dfrac{e^2 + b^2}{e}$ in the equation

$$a^3 - 3b^2 a = 2c^3,$$

if c is greater than b, we reduce it to the simple equation $e^3 = c^3 + \ldots d^3$ (where $c^6 - b^6 = d^6$); if $c = b$, it is reduced to the simple equation $e^3 = c^3$; but if c is less than b, it is reduced to the *impossible* equation $e^3 = c^3 + \sqrt{-d^6}$. Moreover, he concludes by stating : " This third type of equation is impossible since $\sqrt{-d^6}$ is inexplicable " †.

---

* *Algebra*, p. 152.

† " Erit inde $eee = ccc + \sqrt{-dddddd}$, tertii casus æquatio præscripta (propter $\sqrt{-dddddd}$ inexplicabilitatem) impossibilis " (Harriot, *op. cit.*, Sect. 6, Prob. 13).

Further, when he attempts to illustrate this by means of examples, this is what he declares:—

The equation
$$aaa - 6a = 40$$
leads to
$$a = \sqrt[3]{20 + \sqrt{392}} + \sqrt[3]{20 - \sqrt{392}},$$
which reduces to
$$a = 4.$$

But although he has repeatedly told us that "the number of roots is so many as is the index of the highest power", he makes no reference whatever to the other two roots, namely,
$$a = -2 + \sqrt{-6}$$
and
$$a = -2 - \sqrt{-6}.$$

Again, he "solves" the equation
$$aaa - 3a = 52,$$
and he gives as his solution the sole value $a=4$, while the "impossible" values of a, namely, $a = -2 \pm \sqrt{-9}$ are ignored.

Finally, in the equation
$$aaa - 24a = 27,$$
which is reduced to
$$a = \sqrt[3]{36 + \sqrt{784}} + \sqrt[3]{36 - \sqrt{784}},$$
he gives as his solution—and Wallis does not correct him—
$$a = 6,$$
a solution which is inaccurate, as it is incomplete. Such is Harriot's treatment of imaginaries, which, says Wallis (*Algebra*, p. 174), are a "great discovery of Harriot, and wherein Descartes follows him".

Thus it becomes more and more apparent that the pompous parade which Wallis makes of the achievements of Harriot bears but slight resemblance to what is actually displayed in the pages of the *Artis Analyticæ Praxis*, and it is impossible to hold Wallis guiltless of something more than partiality towards its author, who was, if we judge by the *Praxis*, but an indifferent mathematician. Cantor, having in mind Harriot's neglect of negative roots and his "proof" that an equation of the type
$$eee - 3.bbe = -ccc - 2.bbb$$

CHAPTER IX

is *impossible*, is of the opinion that Wallis's eulogies can be explained only by assuming that Wallis was thinking of another work! " Wallis made this mistake in his *Algebra* of 1685 ", says that distinguished historian : " Whoever compares his account with the *Artis Analyticæ Praxis* is forced to the conclusion that Wallis had quite a different work in mind " *.

Cantor's observation suggests that Wallis may have been familiar with other works of Harriot which had not the blemishes of the *Praxis*, and that it was a consideration of these which evoked such commendatory tributes from him. But among the Aubrey collection in the Bodleian Library is a letter from Wallis, written almost on the eve of the publication of the *Algebra*, which dispels this suggestion and renders his homage still more inscrutable. " I am very glad ", wrote Wallis, " to hear tidings of Mr. Harriot's papers (who was a very great man in his time) . . . . *I have never read any of those things, but only that of his Algebra*, which had the good hap to be published by Mr. Walter Warner, as a Prodromus to some other of his works which at that time he gave hopes of publishing, but hath not done it " †. In a subsequent letter he went on to say that to recover these papers would be a very acceptable work. " But I adde also that he would thereby do right to the memory of Mr. Harriot, and indeed to the English nation, who have been very early in the knowledge of these studies, and from whom foreigners have borrowed (to give it no worse a name) very much, without being so kind as to let the world know from whom they had it. But I could wish, that if his Lordship have any special kindness for the memory of this great man (without any prejudice to his own reputation) he would cause them to be made publick in print " ‡. It is doubtful if the papers to which Wallis referred ever existed ; certainly his Lordship (*i. e.*, the Earl of Clarendon) denied all knowledge of their existence.

Nor must it be thought that Harriot was the only Englishman whose mathematical achievements were so conspicuously paraded in the pages of Wallis's *Algebra*. The works of another Englishman, William Oughtred, received from Wallis a measure of praise hardly inferior to that which was lavished upon Harriot. It will be recalled that the study of Oughtred's *Clavis Mathematicæ* had been a powerful stimulus to Wallis in the early days of his mathematical career. Wallis never forgot this. But, as with Harriot, his homage

---

* " Diesem Fehler verfiel J. Wallis in seiner Algebra von 1685. Wer seinen Bericht mit der *Artis Anal. Praxis* vergleicht, muss glauben, Wallis habe ein ganz anderes Werk vor Augen gehabt " (Cantor, ii, p. 792).

† Aubrey MSS. 13, fol. 342, July 20, 1683 (Bodleian).

‡ *Ibid.*, March 8, 1683/4.

towards Oughtred distorted his vision and led him to extol the latter's work also far beyond its merits. For Oughtred was not unindebted to earlier writers, and Gilbert Clark, in his *Oughtredus Explicatus* * (1682) very convincingly points out his obligations to Vieta. But Wallis's views were very different. In a fragment amongst his correspondence in the Bodleian we read: " Mr. Wm. Oughtred follows Vieta (as he did Diophantus) .... But doth abridge Vieta's characters or species only by letters $q.\ c.$ .... which in Vieta are expressed by *Quadrate, Cube*, etc. He doth also to very great advantage make use of several ligatures or compounding notes to signify the sum or difference of several quantities " †. In the Preface to the *Algebra* he classed him with Harriot. " But they were both (Harriot and Oughtred) very great men in this kind of knowledge, and scarce to be equalled (if at all) by any of that age ", and after he had enumerated the achievements of Vieta, he added : " He hath contracted all that Vieta taught into much less room .... The method of Vieta is followed and much improved .... and it [*i. e.*, the *Clavis*] doth in a brief compendious method declare in short what had been before the subject of large volumes." How much reliance is to be placed upon this opinion becomes only too plain when it recalled that where Vieta had shown considerable skill in handling cubic equations, Oughtred is content to deal with nothing higher than quadratics and even then he takes notice of positive roots only. Moreover, in spite of the alleged advantages of Oughtred's notation, Wallis rarely had recourse to it himself and in his earlier writings, *e. g.*, the *Arithmetica Infinitorum*, he was sagacious enough to use the index notation of Descartes. " I would rather explain by means of arithmetical ' degrees ' than by means of geometrical dimensions " he had said thirty years earlier ‡. And no wonder ; the only effect of Oughtred's " several ligatures or compounding notes " is to make the treatise almost unintelligible ! This, by the way, was also the considered opinion of Collins, for in 1667 he wrote : " And as for Mr. Oughtred's method of symbols, this I say to it, it may be proper for you as a commentator to follow it, but divers I know men of inferior rank, that have good skill in Algebra that neither use nor approve it " §.

It therefore becomes more and more inexplicable why Wallis should have entertained such an exalted opinion of the abilities

* Gilbert Clark, pp. 121 and 159.
† Add. MSS. D. 105, Fol. 104 (Bodleian).
‡ " Mallem autem per varios Gradus Arithmeticos, quam per Geometricas Dimensiones, rem explicare " (*Mathesis Universalis*, p. 73).
§ Collins to Wallis (undated) (Rigaud, ii, p. 479).

CHAPTER IX

of Oughtred, for this opinion was certainly not shared by his contemporaries, still less by his successors, in whose estimate Oughtred was by no means an outstanding mathematician. Collins, who was probably more familiar than any other scholar of the century with the mathematical treatises then current, wrote to Wallis : " Mr. Kersey hath made notes on the *Clavis*, and to say the truth, doth not admire anything in it, save what concerns the tenth and succeeding Elements, of Euclid. Mr. Bunning, an aged Minister near Nuneaton in Warwickshire hath commented on the *Clavis* . . . . but one Mr. Anderson, a knowing weaver, told Mr. Bunning that the *Clavis* itself, and his comments thereon, were unmethodical, and the precepts for educing the roots of an adfected equation maim and insufficient " *. Cantor, in his voluminous *Geschichte der Mathematik*, dismisses him in one brief paragraph. Montucla merely remarks : " He developed further the application of analysis to geometrical problems, the construction of equations, the formation of powers, the formulæ for angular sections etc. But the greater part of these things scarcely surpass what one might call elementary analysis, or what we have had already from Vieta. For that reason, it would be useless to tarry longer here " †. We cannot leave this subject, however, without recording the not insignificant fact that Oughtred had on occasions indulged in eulogies of Wallis which were hardly less extravagant. In the Preface to the *Clavis*, for example, he addresses Wallis as " a noble gentleman, devout and industrious, expert in the most abstruse branches of learning ; and particularly acute in mathematics " ‡. Again, in the year following the publication of the *Arithmetica Infinitorum*, Oughtred wrote to Wallis : " I gratulate you, even with admiration, the clearness and perspicacity of your understanding and genius, who have not only gone, but also opened a way into these profoundest mysteries of art, unknown and not thought of by the ancients "§ .

This characteristic, which Wallis so often displayed, namely, an unmistakable tendency to exalt the accomplishments of his

* Collins to Wallis, Feb. 2, 1666/7 (Rigaud, vol. ii, p. 471).

† " Il développa davantage l'application de l'analyse aux problêmes géométriques, la construction des équations, la formation des puissances, les formules pour les sections angulaires &c. Mais la plupart de ces choses ne passent guères ce qu'on pourroit nommer l'analyse élémentaire, ou ce qu'on tenoit déjà de Viète. C'est pourquoi il seroit inutile de nous y arrêter davantage " (Montucla, ii, p. 105).

‡ " Vir ingenuus, pius, industrius, in omni reconditiore literatura versatissimus, in rebus Mathematicis admodum perspicax " (Preface to *Clavis*).

§ Oughtred to Wallis, Aug. 17, 1655 (Rigaud, i, p. 87).

fellow-countrymen to the detriment of equally distinguished continental mathematicians, is greatly to be deplored and must be taken into consideration in any estimate of Wallis's character or of the influence he exerted upon his age. As early as 1667, the diplomatic Collins had tried to suppress this dangerous trait, for in that year he wrote to Wallis : " You do not like those words of Vieta in his theorems *ex adjunctione plano solidi, plus quadrato quadrati* etc. and think Mr. Oughtred the first that abridged those expressions by symbols ; but I dissent and tell you 'twas done before by Cataldus Geysius and Camillus Gloriosus, who in his first decade of exercises (not the first tract), printed at Naples in 1627, which was four years before the first edition of the *Clavis*, proposeth this equation just as I here give it to you, viz.,

$$1ccc + 16qcc + 41qqc - 2304cc - 18364qc$$
$$-133000qq - 54505c + 3728q + 8064\text{N} \text{ æquatur } 4608,$$

finds N the root of it to be 24, and composeth the whole out of it for proof, just as in Mr. Oughtred's symbols and methods. Cataldus on Vieta came out fifteen years before, and I cannot quote that as not having it by me " *. Moreover, when Wallis had written to Collins, complaining of the brevity and obscurity of Descartes' work, Collins reminded him of what its author had declared, namely, that " he hath done the work of an architect, leaving it to carpenters and masons to finish ". Yet in spite of his own vigorous prejudices, Wallis actually accused the French of partiality. " *Mons. Malebranche* ", he declared in his *Algebra*, " hath lately published .... his *Elemens des Mathematiques* ; which is a Collection out of all or most of the Writers of this nature ; especially from Vieta's time downwards. But for the most part, without troubling his Reader with the Names of the Authors where he found those things by him Collected (*except his two countrymen Vieta and Descartes*) " †.

It cannot be too emphatically asserted that Wallis, in appraising at such a low estimate the achievements of the brilliant French mathematical school, was guilty of the gravest indiscretion. This becomes all the more obvious when it is recalled that Descartes and Vieta were by no means the only French mathematicians who were the subjects of his strictures. Other distinguished members of the French school, notably Du Laurens, suffered from Wallis's too pontifical attitude, as is clear from Wallis's correspondence. To declare, as Wallis repeatedly declared, that the work of Descartes was nothing more than an abstract of what Harriot had previously delivered, was a calumny which a wiser than Wallis would never

* Collins to Wallis (undated ; probably 1667) (Rigaud, ii, p. 477/8).
† *Algebra*, p. 214.

CHAPTER IX

have uttered. Not only was it groundless, but by its all too frequent repetition it completely missed its mark. " Fut-il jamais de déclamation aussi aveugle, et autant contredite par l'admiration universelle des géomètres pour l'ouvrage de Descartes ? Elle porte avec elle-même sa réfutation " *. Harriot's treatise, it is true, had appeared six years before the publication of *La Géométrie*. But if to use results which had appeared in an earlier treatise constitutes plagiarism, then few writers of the seventeenth century could escape the charge, least of all Harriot himself. Writers of all ages have borrowed extensively from the works of their predecessors and contemporaries, and many of the discoveries which Wallis attributes to Harriot, notably his treatment of cubic equations, are to be found in the pages of Vieta, whose *In Artem Analyticam Isagoge* appeared as early as 1591. " Mais si Descartes a allumé son flambeau à celui d'Harriot, ce qui peut-être ", observes Montucla, " quoiqu'il soit assez vraisemblable que les découvertes principales de sa Géométrie sont antérieures à la date de l'ouvrage de l'analyste anglois, est-ce que Harriot n'a pas probablement allumé le sien au flambeau de Viète, dont tous les écrits ont été publiés avant 1600 ? " †.

A glance at the works of Vieta clearly establishes the fact that, although he did not recognize negative roots, he did thoroughly understand the relations between the positive roots of an equation and its coefficients, and in Chapter xiv of his treatise *De Emendatione Æquationum* (which was published by Alexander Anderson in 1615, *i. e.*, six years before Harriot died) are to be found many of the improvements which also appear in Harriot, such as the removal of the second term of the cubic; moreover, he states that the equation whose roots are $x=a, b,$ and c is

$$x^3-(a+b+c)x^2+(ab+bc+ca)x=abc.$$

It would be more than an exaggeration to suggest that Harriot ever went further than this; in fact, after comparing the works of the two writers, we find no difficulty in agreeing with the dictum of Cantor that, if we are to judge by the *Artis Analyticæ Praxis*, Harriot can never be regarded as Vieta's rival, but only as his pupil ! ‡.

Again, Wallis claimed that Harriot had shown how to increase or diminish the roots of an equation whilst yet unknown and thus

---

* Montucla, ii, p. 119.
† *Ibid.*, p. 120.
‡ " So besteht kein Zweifel, dass, so weit die *Artis Analyticæ Praxis* allein massgebend bleibt, Harriot nur als Schüler, nirgend als Nebenbuhler Vieta's erscheint " (Cantor, ii, p. 79).

effect the elimination of the second term. But this device was well known to many of the mathematicians of the period, particularly Vieta.

Whether Descartes had ever seen Harriot's treatise or no, we cannot say. The fact that he makes no mention of Harriot is not very helpful, for in the pages of *La Géométrie* there are singularly few references to anyone. He certainly did not lack opportunity of seeing it, for, according to Baillet, he was in England in 1631 making observations on the variation of the compass *. On the other hand, in one of his *Lettres*, addressed to Carcavi, he wrote : " He [Roberval] said that in the formation of your equations you only repeat what has been published as early as 1631, by an Englishman named Harriot, of whom we here have no great knowledge, least of all, I " †. But there was not the slightest reason why he should have helped himself to Harriot's work. He could have got quite as much from his own contemporary, Albert Girard, whose *Invention Nouvelle* (1629) had carried the study of equations a good deal further than Harriot ever did. To take a case in point, Girard found the roots of the equation

$$x^4 = 4x - 3 \ (\text{`` Si } 1\,\textcircled{4}\text{ est esgale à } 4\,\textcircled{1} - 3 \text{ ''})$$

to be

1, 1, $-1+\sqrt{-2}$ and $-1-\sqrt{-2}$,

and he adds : " Notez que le produit des deux derniers est 3 " ‡.

But if Descartes had seen Harriot's treatise and even if he had made use of his results, this can hardly be imputed to him as a fault. For after all, Descartes' treatment of equations was only an offshoot of his main theme, and for one writer to build upon foundations laid by another is by no means uncommon, not only in science, but in every branch of human activity. Did not Newton erect his mighty superstructure upon foundations laid by Wallis ? Nor was Wallis himself guiltless of the charge he had so vigorously laid at the door of Descartes. His treatise on the *Conic Sections* and his *Arithmetica Infinitorum*, for example, give evidence only too clear of their author's indebtedness to the analytical methods of Descartes, and in his method of extracting the cube root of a binomial there was little that was new. Again, of the two tangent

---

* Baillet, p. 240.

† *Lettres de Mr. Descartes*, Paris, 1657–67, tom. iii, no. lxxviii, Résponse à Carcavi, Paris, Sept. 24, 1649. " [Il dit] que dans la formation de vos équations vous ne faites que redire ce qui a esté publié dés l'année 1631 par un Anglois, nommé Hariot, duquel nous n'avons pas icy grande connoissance, du moins moy."

‡ Girard, p. 41.

CHAPTER IX

rules given by Wallis in 1672, one, due to Fermat, was published by Hérigone in his *Cursus Mathematicus* (Paris, 1644), while the other, due to Roberval, was known as early as 1636 and is to be found in the works of Torricelli, published eight years later. Finally, much of his work on equations had appeared in 1629 in the treatise to which we have already referred, namely, the *Invention Nouvelle* of Girard.

These examples, which it would not be difficult to multiply, will suffice to indicate that Wallis, in writing a history of algebra, fell far short of the standard he had set himself. Unmindful of the warning he had uttered in his Preface,—" For there is no man can write so warily but that he may sometimes give opportunity of cavilling to those who seek it " *,—he seems to have deliberately cultivated a most intense suspicion towards the achievements of foreigners, particularly the French, and if we turn from the *Algebra* for a moment to his correspondence, we find that this too is replete with instances of the grossest unfairness towards even the most distinguished members of the French school.

In order, therefore, to understand the attitude which Wallis so persistently adopted, we must try to reconstruct the intellectual atmosphere in which he lived. That the seventeenth century was an era illustrious in the annals of science there can be no doubt, but it must also be borne in mind that it was a period in which scientific advance was attended by a bitterness of disputation difficult for the twentieth-century reader to understand. Men did not seem able to realize that it was possible for two scholars to light upon the same discovery simultaneously. Consequently, charges of plagiarism were all too frequent. There is abundant evidence for this. Torricelli and Viviani, for example, discovered certain properties of the cycloid, and these were published by the former in an Appendix to his works in 1643. Their publication at once aroused the jealousy of Roberval, who addressed to their author a long magisterial letter, in which he claimed those discoveries for himself, and he called upon the immortal gods to bear witness to the fact that he was familiar with them many years before the Italian mathematicians had ever turned their thoughts to the subject. In 1646 therefore, when Torricelli wrote to Cavalieri about one of his discoveries, he expressed grave fears lest knowledge of it should reach the French, who would in that event certainly claim it as their own. Descartes seems to have been obsessed with a like fear, for in his *Géométrie* he deliberately left obscurities, since he foresaw that " certain people, who vaunt that they know everything, would not have failed to say that they knew already all

* *Algebra*, Preface.

that I had written, had I made myself more intelligible ". In our own country Hooke claimed to have anticipated Newton's explanation of planetary motion. It was the consciousness of all this that impelled Oldenburg to strive so earnestly to secure rights of priority to inventors. He proposed " that a proper person might be found out to discover plagiarys, and to assert inventions to their proper authors " *. According to Birch, he also introduced a motion on March 15, 1664/5, that " when any fellow should have a philosophical notion or invention, not yet made out, and desire, that the same sealed up in a box might be deposited with one of the secretaries, till it could be perfected, and so brought to light, this might be allowed for the better securing inventions to their authors " †.

It was this dread of plagiarism that drove mathematicians to be so guarded in communicating their results and to have recourse to artifices. Sometimes they concealed their results in anagrams; not infrequently the results were communicated, but not the methods, a device to which Fermat seems to have been particularly partial. Sometimes problems were set as a challenge to the whole mathematical world. It was a very unsatisfactory state of affairs; but it must be remembered that in those days there were no official scientific journals in which mathematicians could establish their claims. Nor was there a sufficiently brisk demand for mathematical treatises to encourage their publication: Oughtred's *Clavis*, for example, was piloted through the press only with the greatest difficulty and even the *Principia* itself had to be sponsored by Halley. The only means of scientific communication and the diffusion of ideas was private correspondence, and this was frequently a scholar's sole link with the external world. This explains the voluminous correspondence of men like Wallis, Collins, Huygens and others. But the unreliability of this means of disseminating scientific information is at once apparent. Depending as it did to an inordinate extent upon friendly or hostile feeling, personal prejudices and the like, it could never be satisfactory. Still less could it establish priority, and in addition to the quarrels in which Wallis was directly involved, it is to the imperfections of such means of communication that we must to no little extent attribute the disputes between Torricelli and Roberval, Newton and Leibniz, Hooke and Hevelius.

The lack of a reliable medium for the intercommunication of ideas was keenly appreciated by Wallis and more than once he raised his voice loudly in protest against the subterfuges to which

* Weld, i, pp. 329-30.
† Birch, *Hist. Roy. Soc.*, ii, p. 24.

mathematicians had recourse. To Oldenburg he wrote : " You do not tell me who sends these [problems], whether sent to the Society at large, or as a challenge to me by name, nor by whom my answer is desired. The truth is, I am allmost weary of such problems from beyond sea, for the result is comonly but some quarrelling. If they be not solved, they insult ; if they be, they cavil and be angry. At best they take up time and give trouble to no purpose. If he that proposeth them can solve them, let him tell the world how, if he think fit; which in my judgment tends more to the advancement of published knowledge than challenging offers to find out what they have found already " *. On receiving a copy of a treatise from Du Laurens, which contained an Appendix, *Solutio Problematis a D. Wallisio totius Europæ mathematicis Propositi*, he indignantly replied : " The proposing of a Challenge of this nature is a vanity I never was guilty of, and 'tis like enough I never may be " †. But his oft-repeated protests cease to be impressive when it is noted that he did not hesitate to employ such methods himself.

In passing judgment upon Wallis, it must also be borne in mind that he always felt that scant justice was being done to his fellow-countrymen, and that by his exaltation of the English mathematical school he was actuated by a growing uneasiness that the achievements of the English were in danger of being eclipsed. England had just emerged into the ambit of intellectual Europe, and she was rapidly assuming a place in the world of science no less distinguished than that which she had already won in letters. Wallis fully recognized this and he was determined that his country should not be ousted from her hard-won position. Of course this does not excuse the violent polemics with which the pages of the *Algebra* abound, but it may help to explain them. Thus Wallis repeatedly complained that Harriot was little taken notice of abroad ; moreover, in many of his letters, written between 1695 and 1697, he showed the greatest alarm lest Newton's observations in optics might be anticipated by foreigners. Again, in a letter to Boyle, he wrote : " There be two reasons by which you have prevailed with me at last to do something. First because it is the common Fate of the *English*, that out of a modesty they forbear to publish their discoveries till prosecuted to some good degree of certainty and perfection ; yet are not so wary, but that they discourse of them freely enough to one another, and even to strangers upon occasion ; whereby others, who are more hasty and venturous, coming to hear of the notion, presently publish something of it, and would be

* Wallis to Oldenburg, Dec. 10, 1668.　(*Royal Society Guard Books.*)
† *Ibid.*　Wallis to Oldenburg, Mar. 30, 1668.　(*Royal Society Guard Books.*)

reputed thereupon, to be the first inventers thereof : though even that little, which they can then say of it, be perhaps much lesse, and more imperfect, than what the true authors could have published long before " *. To Oldenburg he wrote in the same year : " I could wish that those of our Nation were a little more forward than I find them generally to bee (especially the most considerable) in timely publishing their own discoveries and not let strangers reape the glory of what those amongst ourselves are the authors " †. Two years later, he wrote to Collins : " Having found the quadrature of the hyperbola a little lame (that is, not so full as I wished it were), I did that night consider how it might be improved, and the next morning, I wrote to Mr. Oldenburg because I was unwilling to leave to foreigners the perfecting of that which was by ours carried so far " ‡.

Herein, no doubt, lies the explanation of this unfortunate characteristic of his. Yet one cannot fail to observe that it was particularly against the French that his anger was directed, and for this another explanation has been suggested by Thomson, who records the following in his comprehensive *History of the Royal Society* §. In 1658 Pascal once more turned his attention from his literary and religious pursuits to the study of mathematics. Like many other mathematicians of the period, he became absorbed in the study of the cycloid, and many new properties of the curve were discovered by him. These properties Pascal was persuaded to make use of in order to test the strength of the mathematicians of Europe. Accordingly, under the name of A. Dettonville, he addressed a challenge to all the geometers of Europe, dated June 1658, inviting them to solve his problems and offering a prize for their solution if received before October 1st. The problems, however, were of so difficult a nature that few mathematicians of the age were capable of solving them ; in fact, only two men sent solutions, the Jesuit Father Lalouère of Toulouse and Doctor Wallis. Neither received the prize. Carcavi and Roberval, who were appointed commissioners to examine the papers, decided (November 25, 1658) that Wallis's papers contained several errors. Wallis, however, when he published his tract on the *Cycloid* (1659) denied this. He acknowledged a few inadvertencies, but these, he maintained, were of little moment. The unfortunate contest seemed to bring out the worst side of the disputants, and there

* Wallis to Boyle, April 25, 1666. (*Royal Society Guard Books.*)
† Wallis to Oldenburg, Mar. 21, 1666/7. (*Royal Society Guard Books.*)
‡ Wallis to Collins, July 21, 1668. Rigaud, ii, p. 491.
§ Thomson, p. 269.

CHAPTER IX

can be no doubt that the decision embittered Wallis against the French mathematicians. For when Huygens wrote to him, asking him his opinion of Pascal's treatise, *Histoire de la Roulette*, in which the correct solutions were eventually published, Wallis replied that though the piece were well enough executed, yet there was nothing extraordinary in it considering the great assistance the author had received from the solutions to the prize questions, which, upon recollecting that unhappy affair, ... he could not but be of the opinion it was a mere decoy to produce for France the honour of the invention *. Moreover, it must not be forgotten that his strictures were not confined to Pascal, Descartes and Du Laurens. From the *Commercium Epistolicum* (1657) we learn that Fermat and Frénicle had already been drawn into a violent controversy, a controversy which did credit to none of its participants. Further, Wallis's animosity towards the French was very persistent. In a letter to Sir Samuel Morland he convicted Descartes of plagiarism at least fifty times, and when Baillet tried to gloss this over, Wallis would have none of it, but went on to say that the French were no friends of his nor was this the first piece of French craft he had exposed. Indeed, he seemed to think it inherent in the nature and genius of that nation that they could not let slip an opportunity of exercising their talent for it ! †. Perhaps no better example of the poor opinion that he had systematically cultivated towards the French could be adduced than a letter to Collins, where, speaking of his own *Arithmetica Infinitorum*, he wrote : " But that book is a book which the French, though 'tis like enough they make good use of it, do not desire the world should take notice of " ‡.

We leave the historical section with an observation with which Cantor opens the third volume of his *Geschichte der Mathematik*. " As to the value of this Algebra, as such ", says he, " we shall speak later ; as to its historical section we can only repeat in a fuller form the opinion which we have on occasion already expressed. It is actually not an historical section at all, but rather an example of English partisanship, inspired by excessive national pride. A glorification of Thomas Harriot, of Isaac Newton, of Wallis himself is intended, and in particular the first named has, if one is to believe Wallis, invented everything which is of oustanding importance in the theory of equations .... The worst feature was that the uncritical reader accepted the dogmatic assertions and thus

---

* Wallis to Huygens, Dec. 4, 1659. (Huygens, ii, 518.)
† Letter to Sir Samuel Morland, Jan 8, 1688. *B. B.*, vi, p. 4132.
‡ Wallis to Collins, Sept. 8, 1668. Rigaud, ii, 496.

errors arose, which long remaining undisputed, are handed down from text-book to text-book " *.

But when we leave the historical section, we come to something tangible in Wallis's *Algebra*. Well might young Roger Cotes, himself on the threshold of a brilliant mathematical career, write concerning it : " To my mind, there are many pretty things in yt book worth looking into " †. There Wallis rises to great heights ; every page is impressed with his rich, confident mastery, and arouses a sense of disappointment that such a brilliant and useful piece of work should be marred by a history so aggressively unbalanced. By a strange irony, his best work consists of filling the gaps which Harriot had left, so that the treatise, if anything, calls attention to many of the deficiencies he had laboured so earnestly to conceal. For example, where Harriot is content to discuss equations with positive roots only, Wallis's investigation is far more comprehensive. In his quadratic equations he discusses every type, and the rules he evolves for determining the nature of the roots by a mere inspection of the equation would not be out of place in a modern text-book. He was quite at home with imaginaries, and he knew that such roots always occurred in pairs. Moreover he would not allow the use of the word *Impossible* as applied to an equation with imaginary roots. An equation, for example, such as

$$-aa+8a=25,$$

of which the roots are $4+\sqrt{16-25}$ and $4-\sqrt{16-25}$,

" imaginaries "—had hitherto been styled " Impossible ". " Yet ", avers Wallis, " these are not impossible equations, and are not altogether useless but may be made use of to very good purposes. For they serve not only to show that the case proposed (which resolves itself into such an impossible equation) is an Impossible

---

* " Auf den Werth dieser Algebra als solcher kommen wir später zu reden, über deren geschichtlichen Theil müssen wir das Urtheil ergänzend wiederholen, welches wir gelegentlich aussprachen. Er ist überhaupt kein geschichtlicher Theil, sondern eine von englischem übermässigem Nationalstolze beeinflusste Parteischrift. Eine Verherrlichung von Thomas Harriot, von Isaac Newton, von Wallis selbst ist beabsichtigt, und namentlich der erstgenannte hat, wenn man Wallis glauben schenkt, so ziemlich Alles erfunden, was von hervorragender Wichtigkeit in der Lehre von den Gleichungen ist . . . . Das schlimmste war, dass kritiklose Leser den zuversichtlich ausgesprochenen Behauptungen vertrauten, und so entstanden Irrthümer, welche lange Zeit unangefochten von Lehrbuch zu Lehrbuch sich forterbten " (Cantor, iii, p. 4).

† Edleston, p. 191.

CHAPTER IX

case, and cannot be so performed as was supposed, but it also shows the measure of that impossibility, how far it is impossible, and what alteration in the case proposed would make it possible. And although they are of use in compounding Superior Equations, which though as to these Roots they will be impossible, may as to other Roots be possible Equations. Thus the equation

$$aaa - 7a = 6$$

has for its roots the Imaginary quantities

$$\sqrt[3]{3 + \sqrt{-\frac{100}{27}}} + \sqrt[3]{3 + \sqrt{-\frac{100}{27}}},$$

and is therefore looked upon as an Impossible case. Yet it hath a real Root $a=3$, besides which it hath also two negatives $a=-1$, and $a=-2$ " *.

Again, Harriot had shown how, by multiplication, compound equations could be derived from laterals. But it was Wallis who carried the investigation much further, by illustrating how these compound equations might by division be reduced to much simpler equations. Thus to take Wallis's own example.

The equation
$$aaa - baa + bca - bcd = 0$$
$$-caa + bda$$
$$-daa + cda$$

is composed of three laterals,

$$a - b = 0,$$
$$a - c = 0,$$
$$a - d = 0.$$

Now suppose the compound equation be divided by one of the simples, e. g., by

$$a - d = 0,$$

the result is a quadratic equation containing the other two roots, viz.,

$$aa - ba$$
$$-ca + bc = 0.$$

Hence to solve an equation such as

$$aaa - 10aa + 31a - 30 = 0.$$

* *Algebra*, p. 174.

"If by any means," declares Wallis, "I have discovered the value of one root, (suppose $a=2$) I may (dividing the original equation by $a-2$) depress that cubic into a quadratic, namely,

$$aa - 8a + 15 = 0,$$

which can easily be solved."

Wallis then turns to Descartes' Rule of Signs, by which the number of negative and affirmative roots of an equation may be determined by inspection. "This rule", says Wallis, "is either a mistake or an inadvertence, for it must be taken with this caution, viz., that the Roots are Real not only Imaginary" *. As evidence of this, he proposes the equation:

$$x^4 + 6x^3 + 111x^2 + 1993x + 35878 = 0$$

into $\quad x - 18$

makes $\quad x^5 - 12x^4 + 3x^3 - 5x^2 + 4x - 645804 = 0.$

Now, by the rule the first of these should have four negative roots, and the second one affirmative. And therefore the roots of the third equation, which is compounded of the other two, should be one affirmative and four negatives. Yet, by the same rule, all the five in the compound equation should be affirmative. Thus it is manifest that the rule given by Descartes requires modification.

Harriot had explained the artifice of adding to, or subtracting from, a root whilst it was still unknown, in order to eliminate a term of the equation. This device had been known also to Vieta and to Cardan. But it remained for Wallis to demonstrate how, by this means, the solution of a quadratic reduced to the mere extraction of a root. Thus in the equation

$$aa - 2ba = cc$$

he puts $\qquad a = e + b,$

whence $\qquad ee + 2be + bb = aa$

$\qquad\qquad -2be - 2bb = -2ba,$

which becomes $\qquad ee = bb + cc,$

and $\qquad e = \pm \sqrt{\,: bb + cc},$

one of which values will be found to be affirmative, and the other negative. In this way, Wallis demonstrates the solution of the different types of equations, tabulating his results thus:

Equation.	The Roots are:	
$aa - 2ba = cc.$	$a = +b \pm \sqrt{\,: bb + cc}.$	The greater Affirmative.
$aa + 2ba = cc.$	$a = -b \pm \sqrt{\,: bb + cc}.$	The greater Negative.
$aa - 2ba = -cc.$	$a = +b \pm \sqrt{\,: bb - cc}.$	Both Affirmative.
$aa + 2ba = -cc.$	$a = -b \pm \sqrt{\,: bb - cc}.$	Both Negative.

* *Algebra*, p. 158/9.

CHAPTER IX

From quadratics, Wallis passes to cubics. Harriot, in discussing equations of the third degree, investigates the equation, (12th and 13th Examples)

$$aaa + 3bba = 2ccc.$$

His method is substantially the same as that of Vieta given nearly fifty years earlier. Putting $a = \dfrac{ee-bb}{e}$, the equation becomes, (using the Index Notation—which Wallis does not, any more than does Harriot)

$$\frac{e^6 - 3b^2e^4 + 3b^4e^2 - b^6 + 3b^2e^4 - 3b^4e^2}{e^3} = 2c^3,$$

whence $\qquad e^6 = b^6 + 2c^3e^3,$

which is an affected quadratic in e^3, and which he is able to solve by completing the square *.

In Chapter xlvi Wallis gives a new method of solving the cubic, which he says he discovered about 1647. The method is substantially an application of Cardan's Rules †, though Wallis disclaims having had any knowledge of these at the time. Nevertheless, his investigation includes a method of extracting the cubic root of a Binomial which he claimed as his own. " Now because he (*i. e.*, Harriot) doth not tell us by what method he finds the Binomial Root of his Binomial Cube, I shall set down a method of my own because I have not met with a better, invented many years agoe " ‡. This method, however, is the same as that given more than fifty years before by Girard ; in fact, if Cantor is right, it was much older than that. Wallis illustrates his method by solving the equation

$$aaa - 6a = 40.$$

Putting $\qquad a = \dfrac{e^2+2}{e},$

the equation becomes $\quad e^6 - 40e^3 + 8 = 0,$

whence $\qquad e^3 = 20 \pm \sqrt{392},$

so that Wallis's next task is to extract the cube root of this expression. To do this he first proceeds to " exempt from the note of radicality

* The *Algebra* puts $a = \dfrac{ee+bb}{e}$, obviously a mistake for $\dfrac{ee-bb}{e}$. These errors were not uncommon in the *Algebra*.
† See Note on the Cubic in Appendix I, under *Cardan*.
‡ *Algebra*, p. 175.

so much of it as is rational ", and thus he obtains $14\sqrt{2}$ from $\sqrt{392}$, as Girard had done. "Now", says Wallis, "it is manifest that if this Binomial Cube have a Binomial Root, one part of it must needs be $\sqrt{2}$ or at least some multiple of this by a rational number." From this he easily shows the root to be $2\pm\sqrt{2}$, whence it follows that a, which $=\dfrac{e^2+2}{e}$,

is
$$\dfrac{8\pm 4\sqrt{2}}{2\pm\sqrt{2}}=4.$$

"Therefore", concludes Wallis, "those equations which have been reputed desperate are as truly solved as the others, and thus by casting out the second term, the cubic may be reduced to one of the following forms, of which one root at least is Real, Affirmative or Negative; (the others being sometimes Real and sometimes Imaginary)"*.

Equation.	Root.
$aaa+3ba-2d=0.$	$+\sqrt[3]{d+\sqrt{dd+bbb}}-\sqrt[3]{-d+\sqrt{dd+bbb}}=a.$
$aaa+3ba+2d=0.$	$-\sqrt[3]{d+\sqrt{dd+bbb}}+\sqrt[3]{-d+\sqrt{dd+bbb}}=a.$
$aaa-3ba-2d=0.$	$+\sqrt[3]{d+\sqrt{dd-bbb}}+\sqrt[3]{+d-\sqrt{dd-bbb}}=a.$
$aaa-3ba+2d=0.$	$-\sqrt[3]{d+\sqrt{dd-bbb}}-\sqrt[3]{+d-\sqrt{dd-bbb}}=a.$

Having given the solution of each of these types, Wallis is satisfied that he has given a solution of all cubic equations, at least, as far as one of its roots is concerned. But he has previously shown that once one root is discovered, the cubic can always be reduced to a quadratic by division. Moreover the principle having been demonstrated, that every cubic equation can be reduced to a quadratic, it is an easy step to proceed to equations of a higher degree, and reduce them to a lower degree, first by elimination of the second term, and then by division. This he illustrates by solving the equation

$$a^4+4ba^3=c^4,$$

which he effects by putting $a=e-b$.

With this, Wallis leaves pure Algebra for a brief interval, and in Chapter liii he turns his attention to the solution of certain algebraical problems by the application of geometrical methods. He makes no reference to the work of Descartes in this branch of mathematics, though he refers to "the store of examples in Vieta, Oughtred, and modern writers", and, on the whole, this section

* *Algebra*, p. 279.

CHAPTER IX

falls far below the pioneer work he had done on the solution of equations. In fact, he admits that geometrical applications did not appeal to him, since they lack generality. "Whereas I find", he said, "some others (to make it look, I suppose, the more Geometrical) to affect Lines and Figures, I choose rather, (where such things are accidental), to demonstrate universally from the nature of Proportions, and regular Progressions; because such Arithmetical Demonstrations are more abstract and therefore more universally applicable to particular occasions. Which is one main design I aimed at in my *Arithmetick of Infinities* "*. Harriot, he notes, " meddles not with Geometrical Effections ". It is impossible to avoid the speculation that Wallis's disparagement of " Geometrical Effections " was altogether unrelated to the fact that it was in this branch of mathematics that the French mathematicians, notably Fermat and Descartes, had shown conspicuous ability. In any case, it is not surprising to note that, after this brief digression, he returns to what was to him a more congenial investigation, the solution of equations. In Chapter lv he discussed a Rule of Descartes for dissolving a biquadratic equation into two quadratics, but the chapter, like many another, soon degenerates into little more than a philippic against Descartes. For Descartes, he asserts, seems to have been so well satisfied with these Improvements of Algebra to be found in Harriot, and the condition to which it was by him reduced, that in his *Géométrie*, (first published in January 1637) he doth perfectly follow Harriot almost in everything. And adds very little of his own to what we have shown out of Harriot. Besides this, continues Wallis, Descartes affordeth us some Geometrical Effections and Accommodations of Algebra to divers Geometrical Propositions. As Vieta, Getaldus, Oughtred and many others have done since. " But there is one rule (which I do not find in *Harriot*) for dissolving a Biquadratic Equation (whose second term is wanting) into two Quadraticks by the help of a Cubick Equation of a Plain Root (as *Bombel* and *Vieta* had done before him). And this (so far as I remember) is the only thing which he adds (of pure *Algebra*) to what we have above related out of Harriot. His rule differs not in substance from those of Bombel or Vieta .... How he came by it he doth nowhere tell us nor give us any demonstration of it (perhaps because the same had been before shown in Bombel and Vieta) " †.

Having exhausted the subject of equations, Wallis gives us a much desired relief by turning aside from Harriot. His parting shot, however, constitutes the supreme example of his pro-Harriot

* *Algebra*, p. 292.
† *Ibid.*, p. 208.

prejudices. Having made a scanty reference to other mathematicians of the period, such as Bartholinus, Barrow, van Schooten, Huygens, Newton, James Gregory, he actually adds : " I can hardly expect any new improvements of pure *Algebra* other than what is to be built on the foundations laid by *Harriot*, and assumed by *Des Cartes*"*.

In Chapter lxvi Wallis discusses Negative Squares. It will be recalled that most algebraists of the period had been content to regard " imaginary " quantities as " impossible ", and even useless. Cardan, for example, spoke of them as " fictitious ", and more than once Wallis had protested against the neglect of such quantities. Any negative quantity whatever, says he, even if it is not a square, implies an impossibility, since it is not possible that any quantity should be less than none ! Yet do we on that account reject as useless the supposition of negative quantities ? Just as a definite meaning can be assigned to negative quantities, so also, argues Wallis, can a definite meaning be assigned to Imaginaries. " These Imaginary Quantities (as they are commonly called) arising from the *supposed* root of a negative square (when they happen) are reputed to imply that the case proposed is *Impossible*. But suppose that we gain from the sea 10 acres, but that we lose 20. Our gain must be -10 acres, or -1600 square perches. Now suppose this negative plain, -1600 square perches, to be in the form of a square, must not this supposed square be supposed to have a side ? And if so, what shall this side be ? We cannot say that it is 40 or -40, but rather it is $\sqrt{-1600}$, or $40\sqrt{-1}$, where $\sqrt{}$ signifies a mean proportional between a positive and a negative quantity " †. He seems to have hit upon a like notion a dozen years earlier, for in 1673 he wrote to Collins : " If we suppose such a negative square, we may as well suppose it to have a side, not indeed an affirmative or a negative length, but a supposed mean proportional between a negative and a positive, thus designable $\sqrt{-n}$, or rather $\sqrt{-n^2}$, that is, $\sqrt{+n \times -n}$, a mean proportional between $+n$ and $-n$. Only, though I had from the first a good mind to do it, I durst not without a precedent, when I was so young an algebraist . . . . take upon me to introduce a new way of notation, which I did not know of any to have used before me " ‡. Thus he reached a position where he would be supposed to draw a line perpendicular to the real axis, and say this might be taken as the real axis, but he only touched lightly upon the possibility, and his efforts to represent complex quantities graphically fell just short of success.

* *Algebra*, p. 213.
† *Ibid.*, Chapter lxvii, pp. 264–6.
‡ Wallis to Collins, May 6, 1673. (Rigaud, p. ii, 577/8).

CHAPTER IX

In Chapter lxvii he demonstrates the geometrical solution of cubic and quadratic equations, and he showed that the solution of such equations required the construction of a Conic. Much of this ground had, however, been traversed by Descartes and Vieta, though even in this he refused to acknowledge the genius of the former. For having referred to a somewhat obscure work, Baker's *Clavis Geometrica* *, he actually adds: " That of Descartes I shall not need here to repeat, because it is contained (for the substance) in that of Mr. Baker, *as being but a particular case of Baker's general* " †.

In Chapter lxxiii he refers to the subject which had always made the greatest impression upon him, namely, the subject of Infinitesimals. He gives a concise account of what he had already taught in the *Arithmetica Infinitorum*, together with a summary of the work which had been done by others, notably by Newton, subsequent to the publication of that work, nearly thirty years earlier. This is probably the most valuable part of the work, and shows to what heights he was capable of rising when he kept his judgment within bounds and did not allow it to become unbalanced by such vigorous prejudices. His work, however, on that subject has already been discussed at length.

One other aspect of Wallis's work is worthy of notice, namely, his reliance on the Method of Induction, which he says is " plain, obvious and easy, and where things proceed in a clear and regular order (as here they do), and shews the true natural investigation, Which to me is much more grateful and agreeable than the Operose Apogogical Demonstrations (by reducing them to Absurdities and impossibilities) ..., and if any think them less valuable because not set forth with the pompous ostentation of Lines and Figures, I am quite of another mind .... For where the truth of a Proposition depends merely upon the nature of Number or Proportion .... it is much more natural to prove it abstractly from the nature of Number and Proportion without such embarrassing the Demonstration .... And I look upon this as the great Advantage of Algebra, that it manageth Proportions abstractly, and not as restrained to Lines, Figures or any such particular subject ; yet so as to be applicable to any of these particulars as there is occasion " ‡. It will be remembered it was largely on account of this aspect of his demonstrations that the *Arithmetica Infinitorum* had met with no little hostile criticism, particularly from Fermat. In the *Algebra* (Chapter lxxviii) Wallis enlarged upon the replies he had already

* *Thomas Baker*: *Geometrical Key,* or the Gate of Equations Unlocked. English and Latin, 1684.
† *Algebra*, p. 273.
‡ *Ibid.*, p. 298/9.

made to Fermat, and his views on the Method of Induction were almost two centuries in advance of his time. He emphasized the fact that his method is a good method of *Investigation* " as that which doth very often lead to the easy discovery of a general rule, or at least a good preparative to such an one " *, none the less valuable because " not set forth with the pompous ostentation of lines and figures " †. Moreover, as to Fermat's censure that the demonstration could have been given in lesser room, Wallis declares : " I permit it to the Reader's judgment whether any one of those three demonstrations of this Rectification, (the longest of which doth not extend to a quarter of a sheet, and all three to a little more than half a sheet,) be not as clear and satisfactory to the understanding of any indifferent mathematician as his long process of more than five sheets of paper. And if so, I know no reason why he should disparage those shorter methods in comparison of such his prolix process . . . . For my own part, I take a Demonstration, if clear and cogent, to be the better for being short. Sure I am that since the introduction of such methods, Mathematicks have been much more improved in this present age than they had been in many ages before. Such things as were wont to be looked upon as profound discoveries are by general methods of which I take the Arithmetick of Infinities to be none the most contemptible, easily discoverable by a direct calculation ‡. By his directing attention to such generalized methods Wallis has deserved well of those who followed him. Most of the mathematicians who preceded Newton had been content with particular solutions ; generalized methods were strangely lacking. Wallis, however, adopted a completely different outlook. After solving one of the many problems submitted to him by Fermat, and having given a general solution, he wrote : " But that which I aim at, in discovering these Methods, is not so much for this one Question (which perhaps may not deserve it) as to give a Pattern how other Numeral Questions of like nature (or even more perplexed than this) may in like manner be solved by continual approaches till we come to a coincidence even without an Infinite Process " §. This he illustrates in his solution of other problems set by Fermat, and with this the book closes. With the *Algebra* were also published his *Treatise of Angular Sections*, his *Defense of the Treatise of the Angle of Contact*, and a *Treatise of Combinations, Alternations and of Aliquot Parts*. In none of his writings do we find any definite

* *Algebra*, p. 306.
† *Ibid.*
‡ *Ibid.*, p. 297/8.
§ *Ibid.*, p. 371.

CHAPTER IX

contribution to Trigonometry; it is perhaps for this reason that there was published with the *Algebra* a treatise with the title *A Brief (but Full) Account of the Doctrine of Trigonometry, both Plain and Spherical*, by John Caswell, M.A. (London, 1685).

But it was the *Algebra* which was most widely read despite its blemishes, and so far as his work on equations went, it constituted a reservoir from which contemporary and later algebraists drew much inspiration. It may be mentioned that this treatise did a great deal towards popularizing the notation which was now rapidly becoming current in Europe. He abandoned, as we have shown, the geometrical symbolism, " so that now we need not be frighted at the uncouth names of squared-square, sursolid, as importing more local Dimensions than Nature can admit, for these hard names are but bug-bears " *. He avoided the excessive symbolism of Oughtred, and in many respects he improved upon Vieta, Harriot, and even Descartes. But for some obscure reason he almost completely avoided the Index Notation, for which he had shown such a decided preference thirty years earlier in the *Arithmetica Infinitorum*, and he fell back on the clumsy notation of Harriot. Undeniably, this is a fact greatly to be deplored, since the employment of the Cartesian notation would have materially shortened his demonstrations on the solution of equations, besides adding considerably to their clarity †.

---

* *Algebra*, p. 91.

† Especially is this the case in the English edition of the *Algebra*, which abounds in errors. Thus we frequently meet *eeeee* for e^4, *bbbbb* for b^6, and so on. See *Algebra*, p. 188, for example.

CHAPTER X.

THE DISPUTE WITH HOBBES SUMMARISED.

AN account of the mathematical work of John Wallis can hardly be considered complete without a reference, however brief, to that long, profitless quarrel in which he engaged with Thomas Hobbes, a quarrel which dragged its weary length over a quarter of a century, and which was marked by a strange lack of restraint on the part of each of the disputants. This is a convenient point at which to summarise this deplorable affair.

Before the *Arithmetica Infinitorum* had left the press there appeared the *Elementorum Philosophiæ Sectio Prima*; *De Corpore*, Pars iv, (1655), by Thomas Hobbes, in which the author claimed to give an absolute quadrature of the circle. Wallis at once rushed into the controversial arena. He refuted Hobbes' fantastic claim with his *Elenchus Geometriæ Hobbianæ* (Oxon, 1655), a Latin tract written with no little asperity. Hobbes at once realized that his mathematical reputation was in danger of being jeopardized. He replied to "the insolent, injurious, clownish language of the *Elenchus*"* by publishing his book in English with the addition of what he styled *Six Lessons to the Professors of Mathematics of the Institution of Sir Henry Savile* (1656). In this, after having reasserted his views on the principles of geometry in opposition to those of Euclid, he proceeded to repel Wallis's attacks with no lack of dialectical skill, yet with no less impatience than had characterized Wallis's outburst. Moreover, in the heat of the controversy, he did not hesitate to assume the offensive by a criticism of other works of Wallis, and in his *Lessons on Manners to the two Professors of Mathematics at Oxford*, with which the book closed, he clearly showed that in the matter of invective at least he was the peer of Wallis.

That two such able men as these should have become involved in a dispute which marred the best years of their lives will always remain a matter of wonder and of regret. *Tantæne animis cœlestibus iræ*! For what conceivable good could accrue from it! Wallis was a mathematician of repute. Hobbes was not. His criticism of the *Arithmetica Infinitorum*, which he characterised as "a scurvy book, worthy to be gilded, but not with gold" †, was

* Molesworth, iv, 439.
† *Ibid.*, vii, 301.

but an attempt to hide his inability to understand the current methods of analysis under a cloak of abuse. Unconscious of his own shortcomings, yet anxious to be accounted a pioneer, he had exhausted himself in vain attempts to solve impossible problems. Consequently he was easy sport for Wallis, whose mathematical genius was now generally recognized. But what credit Wallis could achieve, or how his reputation could have been enhanced by shattering the arguments of one so ill-equipped mathematically as Hobbes is not easy to understand. Besides, the dispute was fast becoming more and more acrimonious; it would therefore have been a much more dignified gesture on the part of Wallis to have treated this last outburst with silence. Instead, however, he seemed resolved to out-Herod Herod, for with almost incredible swiftness he paraded Hobbes' mathematical blunders in his *Due Correction for Mr. Hobbes, or School Discipline for not saying his Lessons Aright* (Oxford, 1656), which was even more censorious and unrestrained than his earlier attacks. Hobbes' rejoinder was his Στιγμαι or *The Marks of the Absurd Geometry, Rural Language etc. of Doctor Wallis*, (London, 1657), in which he repelled with vigour, yet not without dignity, the insults which had been loaded upon him. But he could not leave the arena without making political insinuations, which were entirely irrelevant, and which only served to show to what extent the main issue had by this time become obscured. Wallis deftly and easily parried the thrusts in his *Hobbiani Puncti Dispunctio* (1657), in which the mathematical questions were still further relegated to the background. To this Hobbes made no reply.

The publication of Wallis's *Mathesis Universalis* in 1657 was the signal for a renewal of hostilities. Hardly had this appeared in print when Hobbes published a Latin criticism: *Examinatio et Emendatio Mathematicæ Hodiernæ. Distributa in Sex Dialogos* * (Lond., 1660). Wallis very wisely refrained from answering, and his silence provoked Hobbes to resort to subterfuge. Having solved, as he imagined, another of the great problems of antiquity, namely the Duplication of the Cube, he caused his " solution " to be published anonymously in Paris, so as to throw Wallis and other critics off the scent, and so extort a reply which might otherwise be withheld. The ruse was successful. Wallis poured scorn upon Hobbes' paltry effort, which was then promptly acknowledged by its author. About this time Hobbes wrote a severe criticism of Robert Boyle's *New Experiments Physico-Mechanical, Touching the Spring of the Air and its Effects*, in his

* "The Examination and Emendation of Modern Mathematics, collected together in Six Dialogues."

Dialogus Physicus sive De Natura Aeris (Lond., 1661). To his fulminations Boyle replied with his usual quiet dignity, but from his old enemy retribution came in the scathing satire *Hobbius Heautontimorumenos* or *A Consideration of Mr. Hobbes, His Dialogues* (Oxford, Feb. 1661/2). Professing to be roused by the attack on his friend Boyle, when he had scorned to lift a finger in defence of himself, Wallis in this exacted a crushing revenge. He stigmatised his adversary as " a person extreamly *Passionate and Peevish*, and wholly *Impatient of Contradiction* . . . . one highly *Opinionative and Magisterial, Fanciful* in his Conceptions, and deeply *Enamoured with those Phantasmes* without a Rival. He would be thought, of All that *are*, or even *have been*, the only *knowing Man* "*. Not content with invective such as this, he actually charged Hobbes with having written *Leviathan* in support of Cromwell, thus having deserted the Royal cause in its distress. Though overwhelmed by the violence of the attack, Hobbes replied with *Considerations upon the Reputation, Loyalty, Manners, Religion of Mr. Thomas Hobbes*, (1662), a treatise written in the third person throughout. This, which is of great biographical interest, had the effect of silencing Wallis for a time. In 1666, however, Hobbes opened another period of controversial activity, which continued until his ninetieth year. In that year he published his *De Principiis et Ratiocinatione Geometrarum, contra Falsum Professorum Geometriæ* † (1666), which was designed, as the title declared, as a blow to the pride of professors of geometry by showing that there was no less uncertainty and error in their works than there was in those of writers on Physics or Ethics. In the Preface to this treatise Hobbes declared that he invades the whole nation of geometers ; he will acknowledge no judge of this age, for he is confident posterity will pronounce for him. Meanwhile he ventures to advance this dilemma : " Eorum qui de iisdem rebus mecum ediderunt, aut solus insanio, aut solus non insanio Ego ; tertium enim non est, nisi, (quod decet forte aliquis), insaniamus omnes ". (Of those who with me have written something about these matters, either I alone am mad, or I alone am not mad. No third opinion can be maintained unless (as perchance it may seem to some) we are all mad.) " This ", asserted Wallis, " is but a new book of old matter, containing but a repetition of what he hath told us more than once . . . . Moreover, it needs no refutation. For if he be mad, he is not likely

* Heauton-timorumenous. An Epistolary Discourse addressed to Boyle, Feb. 20, 1661/2, p. 3.
† " Of the Principles and Reasons of Geometries, against the Vanity of Professors of Geometry."

CHAPTER X

to be convinced by reason ; on the other hand, if we be mad, we are in no position to attempt it ".

In 1671 Hobbes worked up his " proofs " again, and published them in his *Rosetum Geometricum, sive Propositiones Aliquot frustra antehac tentatæ, cum Censura brevi Doctrinæ Wallisianæ De Motu* * (1671). In this Hobbes persisted in defending his earlier imbecilities as obstinately as ever, and at the same time he displayed the greatest animosity towards Wallis. In the " Brief Censure " he declared that Wallis's treatise *De Motu* abounded in " vicious definitions and pretended obscurities ", and, among other charges, he accused Wallis of using the expression " Infinity " very carelessly. " Can there be supposed ", he asks, " an infinite row of quantities, whereof the last can be given ? .... Is there any quantity greater than Infinite ? " Unfortunately, Hobbes made a number of mathematical blunders in this treatise, and for these he was mercilessly pilloried by Wallis. " I cannot but take notice of his usual trade of contradicting himself ", he said, " I am ashamed that so great a pretender to such high things in Geometry, should be so miserably ignorant of the common operations of practical arithmetic " †.

Still undaunted, the indefatigable Hobbes published a treatise with the high sounding title *Lux Mathematica. Excussa Collisionibus Johannis Wallisii Theol. Doctoris, Geometriæ in celeberrima Academia Oxoniensi Professoris Publici, et Thomæ Hobbesii Malmsburiensis. Multis et fulgentissimis aucta radiis Authore R.R.* ‡ (Lond., 1672). This treatise contains a history of the disputes between the two adversaries. Written with marked perspicacity and vivacity, and in the third person throughout, it gives Hobbes the victory at every point over his antagonist. Later in the year Hobbes requested Oldenburg to lay before the Society a severe criticism of Wallis's works, and this drew forth a rebuke from Oldenburg. " I could not read your letter of Nov. 26 before ye Royal Society, (as by ye Bearer's intimation you seemed to desire me to doe) it being very improper in my judgment to expose so worthy a member as Dr. Wallis is, to that Body by reading publicly an Invective against

---

* " The Geometrical Rose Garden, or Some Propositions hitherto unsuccessfully attempted, with a Brief Censure of the Wallisian Doctrine De Motu."

† *Phil. Trans.*, no. 75, 1671.

‡ " Mathematical Light, struck out from the Clashes between John Wallis, Doctor of Theology, and Public Professor of Geometry in the famous Academy of Oxford, and Thomas Hobbes of Malmesbury, augmented with many and shining Rays, by the Author R.R."

him "*. Wallis's reply to the *Rosetum Geometricum* was sent to the Royal Society, and was subsequently printed in the *Transactions* †. The letters R.R., observes Wallis, denote *Roseti Repertor* (Finder of the *Rose Garden*); it will likewise be answered by R.R.—*Refutator Roseti*, (Refutator of the *Rose Garden*). Hobbes' final shot was his *Decameron Physiologicum* (1678). He died next year, and thus the curtain fell upon the strangest warfare in which able though perverse thinkers ever engaged.

For nearly a quarter of a century the two disputants had waged a contest which shed a lustre round neither of them, and one cannot help wondering why Wallis should have been so eager to expose the mathematical short-comings of one so ill-equipped as Hobbes. For apart from his published exposures, Wallis disparages, and in no unmeasured language, the claims of his antagonist in more than a score of lengthy letters to different members of the Royal Society. Hobbes' claims were so preposterous that a wiser than Wallis would have left him to work out his own destruction. After all, Newton was in his prime when the dispute was at its height; so was Wren. Yet they did not plunge headlong into the arena. Even the incautious Hooke could find better employment.

A possible motive for Wallis's ceaseless attacks upon Hobbes is suggested by Doctor Gregory in the account of Wallis's Life to which we have already referred. " He wrote much against Mr. Hobbes's mistakes and errors in Geometry, and the rather because he hoped that the discovery of them would lessen Mr. Hobbes's credit in his other writings, which he was sensible was at that time too great, considering their influence " ‡. Oldenburg stated that Hobbes " vapoured of his ability to thinking, and declared that Doctor Wallis had no Geometry but what he got from him ", which, continued Oldenburg, was " so intolerable a piece of impudence that he ought to be soundly taught for it " §. Wallis himself wrote, as if in excuse, to Huygens : " Our Leviathan is furiously attacking and destroying our Universities (and not only ours but all) and especially ministers and the Clergy and all religion, as though the Christian world had no sound knowledge . . . . and as though men could not understand religion if they did not understand Philosophy, nor Philosophy unless they knew Mathematics. Hence it seemed necessary that some mathematician should show him by the reverse process of reasoning, how little

* Oldenburg to Hobbes, Dec. 30, 1672. (*Royal Society Guard Books*, W. ii. 4.)
† *Phil. Trans.*, no. 87, p. 5067, 1672.
‡ Gregory, Life of Wallis (Smith MS. 31, Bodleian).
§ To Oldenburg, MS. Savile, 104, 108, Aug. 5, 1671 (Bodleian).

he understands the Mathematics from which he takes his courage ; nor should we be deterred from doing this by his arrogance which we know will vomit poisonous filth against us " *.

But whatever the motive, Wallis does not emerge from the contest covered with glory. The fact that he persisted in his animosity instead of allowing Hobbes to enjoy the twilight of his life in tranquility, certainly weakens our respect for Wallis, more especially as this was by no means the only sea of trouble upon which he launched his barque. Indeed, the frequency of the controversies in which he found himself involved leaves us wondering how his many-sided genius found time to express itself as it did. Had he followed the example of Newton, who found " philosophy such an impertinently litigious lady, that a man might as good be engaged in lawsuits as have to do with her ", and who " valued friends more than mathematical inventions ", posterity would have accorded him a much greater meed of respect. For it must not be forgotten that in the struggles of those early days it was difficult enough to avoid dissensions ; no excuse can be offered therefore for scouring, like a stormy petrel, the wide ocean of controversy. Consequently, our admiration for Wallis is not a little restrained throughout this unhappy affair, especially as his action in not publishing in his *Opera* his philippics against Hobbes was inspired by no noble motive. Although he says he has no wish to trample on the ashes of the dead, he hastens to inform his readers that he was induced to expose the bad reasoning of that pseudo-geometrician at the time of his writing, since by the advantage of his mathematical reputation he found means of spreading the poison of his principles in religion ! For on the threshold of the *Opera* we encounter the statement : " Certain writings against Thomas Hobbes, pseudo-geometer will not be found here, lest it seem that I wish to triumph over one now dead ; although indeed it was necessary to publish them in the past, as matters then were, when he had paraded himself as a great geometer, and in that guise

* " Sed id agebatur a Leviathano nostro, (ut ex scriptis suis, præsertim Anglicanis, facile est colligere) ut toto impetu tum in Academias nostras, (nec nostras tantum sed et universim omnes tum veteres tum recentiores,) tum Ministros præsertim totumque ministerium, et religionem quidem universam incurrat et pessundet penitus ; Quasi nihil sanum nihil non ridiculum vel in Philosophia vel in Religione noverit Orbis Christianus ; et quidem Religionem non intellexerint quia non Philosophiam nec Philosophiam quia non Mathesin. Necessum itaque visum est, tum ut, itinere retrogado, quam parum ipse Mathesin intellexerit, (unde animos sumsit) demonstret saltem aliquis Mathematicus, tum ut id ita fiat ut ne fastu suo terreamur, quem virus omne et spurcitiem evomiturum novimus." Huygens, *Œuvres*, ii, 296, Jan. 1, 1659.

had dared to offer false suggestions about religion to our unsuspecting youth " *.

In conclusion it may be noted that, bitter as this quarrel was, it was but the reflection of the spirit of the age. The pertinacity and violence of attacks upon particular points in mathematics during the period under review seems to have become a kind of disease, or centre of extraordinary virulence. Mathematics is so fundamental and so abstract that it reaches to the very foundations of thought, and seems to provoke the most unaccountable reactions in the minds of great thinkers. The classical example, and the one which will most readily occur to the reader, is Berkeley's criticism of Newton's Fluxions—" the ghosts of departed quantities ". But it is probably not too much to say that relics of this attitude towards mathematics are not lacking even in this the twentieth century.

---

* " Opuscula quædam contra *Thomam Hobbes*, (Pseudo-geometram) olim scripta, non hic habentur ; ne velle videar, de homine jam demortuo triumphare : Quamvis enim prout tunc res erant, id omnino videbatur faciendum (quando sub prætextu Magni Geometræ, qualem se venditabat, ausus est, in Religionis negotio, incautis adolescentibus perperam sentiendi materiam subministrare ;) ne tamen in Geometriæ damnum id cedat, non jam videtur metuendum." Preface to *Opera*.

CHAPTER XI.

Conclusion. Importance of Wallis's Work to
succeeding Generations.

WITH the publication of the *Algebra* Wallis's mathematical activities slackened considerably. His correspondence with the parent society in London now grew more fitful, and from 1686 to 1693, in spite of many efforts to revive it, it ceased entirely. This is perhaps understandable. Wallis had now passed the allotted span of three score and ten; moreover, his ecclesiastical duties continued to absorb a great deal of his time. Brewster's judgment concerning Newton, that had he not been distinguished as a mathematician and a natural philosopher, he would have enjoyed a high reputation as a theologian, might apply with equal force to Wallis. His theological writings were published in 1691, and a cursory glance at these shows that his interest in Divinity never waned, even when his mathematical researches were most intense.

About this time, the Curators of the University Press at Oxford began to collect his mathematical works. These were eventually published in Latin, in three volumes, with a Dedication to William III, at the Sheldonian Theatre, Oxford, during the years 1693 to 1699. The second volume (1693) is noteworthy as containing Sir Isaac Newton's first published account of the invention of the Fluxional Calculus, together with the unfortunate notation which he had adopted *. Not unexpectedly, his *ex parte* account of this discovery at once aroused some hostile criticism, especially in the Leipsic Acts, whereupon not merely did Wallis re-assert Newton's right to the discovery of the method of Fluxions, but he actually claimed that many of Leibniz' alleged improvements were contained in the pages of the *Arithmetica Infinitorum*. In the third volume (1699), was Flamsteed's statement which we have already mentioned regarding an ostensible parallax of the Pole Star, " a noble invention if you make it out ". Wallis's letters to Flamsteed and to Waller between 1695 and 1698 show quite clearly that he believed Flamsteed to have achieved this, from which one can safely conclude that

* " Sint v, x, y, z, fluentes quantitates, et earum fluxiones his notis $\dot{v}, \dot{x}, \dot{y}, \dot{z}$, designabuntur respective. . . . Qua ratione $\dot{v}$ est fluxio quantitatis v, and $\ddot{v}$ fluxio ipsius $\dot{v}$, and $\dddot{v}$ fluxio ipsius $\ddot{v}$. (*Opera*, ii, p. 392.)

Wallis's knowledge of astronomy was not competent. With the publication of this, Wallis bade farewell to the Press, being then well over four score years of age. But before he laid down his pen, he reviewed the work of the century which was now drawing to a close, in the following characteristic passage :—

" Doubtless in the present age which is now drawing to a period, knowledge of all kinds has experienced very great and even unhoped for improvements, as physics, medicine, chemistry, anatomy, botany, mathematics, geometry, analytics, astronomy, geography, navigation, mechanics, and even, what I least rejoice at, the art of war, and indeed far greater than for many years before. For formerly men seemed to aim at nothing more than to understand what had been delivered by Euclid, Aristotle and the rest of the ancients, having little concern about making any further progress as if they had established the limits of the sciences which it might be presumptuous to extend. But after some few had ventured to look further, others were thence encouraged to enter the wide field of the sciences. Thus a new ardour, a new effort urged them to attempt new things and indeed not without success. But when the novelty ceased, this new ardour gradually declined. Many of the diligent scrutinizers of nature are dead . . . . the matter also itself was great, but is now partly exhausted ; so that a harvest is not to be expected, but only a gleaning ; and it seems reasonable to allow that those that are tired and wearied should have some rest. You wish, and so do I that a new ardour be infused into our own Royal Society " *.

But even in his old age his inventive powers never seem to have lost their virility. " This good old gentleman ", wrote Doctor Charlett to Pepys in 1699, " is now as fresh and vigorous for any new undertaking (of any sort) as if he had never put pen to paper " †. But in the same year Wallis was already beginning to complain of " decays ", adding that " 83 is an incurable distemper ‡ ". He died on the 28th October, 1703, at the advanced age of eighty-seven, and was interred in the University Church, where a monument to him was erected by his son John Wallis.

A career of such intense activity, which lasted almost from the cradle to the grave, indicates a vigorous constitution and the possession of a remarkable intellect. " Till I was past four score years of age ", he wrote to Pepys in 1701, " I could pretty well bear up under the weight of those years ; but since that time it

* To Leibniz, April 20, 1699. *Phil. Trans.*, Abridged (Hutton), vol. iv, p. 414.
† *Correspondence of Pepys* (Tanner), vol. i, p. 171.
‡ *Ibid.*, p. 175.

CHAPTER XI

hath been too late to dissemble my being an old man. My sight, my hearing, my strength are not as they were wont to be " *.

Few men of even that illustrious age had a more eventful career. Being in financially easy circumstances he was able to cultivate a wide variety of interests, and in many of these the knowledge he acquired was profound. He lived in a very turbulent period, a period which included the Civil War, the Restoration of Charles II, the Great Plague and the Great Fire, the Revolution of 1688, not to mention the vast political disturbances which were taking place on the Continent. Fortunately for science he was able to rise above the troubled times in which he lived, and some of his best work was produced when his country was facing the gravest crises in her history.

Yet Wallis's career, brilliant though it was, was chequered by disputes of every kind. It was a melancholy fate which ordained that the best years of such an outstanding genius should be distracted and embittered by the fury of opposing principles and the venom of personal abuse. Yet so it was, and it is difficult to understand why Wallis should have been so eager to acquire so great a share of controversial notoriety. Looking back on these disputes dispassionately after the lapse of so many years, and when personal animosities have long been allayed, it is impossible to hold Wallis guiltless. In his letter to Doctor Smith we read : " It hath been my lot to live in a time wherein there have been many and great alterations. It hath been my endeavour all along to act by moderate principles between the extremities on either hand, in a moderate compliance with the Powers in being in those places where it hath been my lot to live, without the fierce and violent animosities usual in such cases against all that did not act just as I did, knowing that there were many worthy persons engaged on either side. And willing, (whatever side was uppermost) to promote (as I was able) any good design for the true interest of Religion and learning for the publick good, and ready so to do good offices as there was opportunity. And if things could not be just as I could wish, to make the best of what is " †. Wallis was more than four score years when he wrote this, and his judgment had no doubt become matured with advancing years. But his whole career shows quite unmistakably that he knew his place as a scholar, and that he was determined at all costs to maintain it and to vindicate his rights, and with them, the rights of his fellow-countrymen. Unhappily his zeal often outran his prudence, and led him as we have seen, to hurl the basest of charges against some who were intellectually his equals, if not definitely

* Pepys, iv, p. 365, Sept. 24, 1701.
† Hearne, p. 169.

his superiors. This is all the more to be deplored when it is recalled that, if Aubrey is right, Wallis often laid himself open to charges similar to those which he did not hesitate to hurl at others. For that well-known gossip did not hesitate to censure him. " 'Tis certain y^t he is a person of reall worth ", he wrote, " and may stand with much glory on his own basis, and need not be beholding to any man for fame, of which he is so extremely greedy, that he steales feathers from others to adorne his owne cap—e. g., he lies at watch, at Sir Christopher Wren's discourse, Mr. Rob. Hooke, Dr. William Holder etc., putts downe their notions in his note booke, and then prints it without owneing the authors. This frequently of which they complaine " *. Moreover, Wallis's eagerness to pose as a *censor morum*, together with his reluctance to acknowledge the genius of foreigners, particularly the French, was probably not without some influence upon that lamentable alienation of the English and the Continental Schools which was such a disquieting feature of the early eighteenth century. It must come as no surprise, for example, to find that when Biot gave an account in the *Commercium Epistolicum* of those whose labours had prepared the way for the discovery of the Infinitesimal Calculus (*l'Analyse Infinitésmale*) he entirely omitted the name of Wallis, although writers who had only remotely contributed to the discovery, as for example Hudde, Slusius, are mentioned. Further (p. 254), he later excused the omission by saying : " On s'étonnera peut-être, de ne recontrer ni Wallis ni Huyghens.... Cependant Wallis et Huyghens ne me paraissent avoir aucun droit direct de paternité sur les nouveaux calculs, qu'ils ont tous deux méconnus, le premier plus encore peut-être que le second " †. In the same way, de Gua in his account of recent development in algebra, published in the *Memoirs of the Academy of Science*, makes nearly all the work of Englishmen ancillary to that of Vieta. It is difficult not to believe that this was to a large extent merely a repercussion of the attitude adopted by Wallis.

In the Manuscript *Life of Wallis* by Lewis we read that Wallis was endowed with " a hale and vigorous constitution of body, and a mind that was strong, serene, calm and not soon ruffled " ‡. The latter part of this is certainly not true ; nevertheless it would be a grave injustice to the memory of this remarkable man to suggest that this unhappy characteristic, namely his passion for controversy, had not some deep-seated cause. It must be remembered that

* Aubrey, II, ii, p. 570.
† *Commercium Epistolicum*, J. Collins et Aliorum. Biot et Lefort (1856), p. 254.
‡ Additional MSS. 32061 (Bodleian).

Wallis lived in a very turbulent age. Moreover he had the vexation of seeing his two rivals, Ward and Wilkins, neither of whom was more acceptable to the Government than he, raised to the highest dignity in the Church, whilst he was repeatedly passed over. Again, the fact that he was the master of a very dangerous art had made for him many enemies. Henry Stubbe, for example, in his treatise *A Severe Inquiry into the late Oneirocritica* (1657) tells us that " the Doctor hath decyphered (besides others, to the ruin of many loyal persons) the King's Cabinet taken at Naseby, and as a monument to his noble performance, deposited the Originals with the decyphering with the Public Library at Oxford " *. Wallis stoutly refuted this charge on several occasions. To John Fell, for example, he wrote regarding the Papers taken at Naseby Fight : " The thing is quite otherwise. Of those letters and papers (whatever they were) I never saw any one of them but in print, nor did those papers as I have been told need any deciphering at all either by me or by any one else " †. Nevertheless, on close scrutiny, it seems fairly clear that Wallis did on occasions exercise his dangerous skill incautiously. Moreover, it is not without significance that Wallis's grandson, William Blencow, learned this art,—from whom it is impossible to say,—and the following letter, found amongst Wallis's private correspondence, is not insignificant.

" My father ", wrote Blencow, " thought it proper that I should wait upon Baron Spanheim with two of the letters I had decyphered to-day, as having orders from you so to do, and told him that if he thought it would answer the pains that must be taken about a business of this nature, I would endeavour to decypher the other two. I found by his answer that he was so disgusted at your letters, and had, I believe, no thought of answering them, he said further that he had received a letter from his King yesterday by which he found out that all opportunity of making advantage of these was lost, and therefore it would not be necessary for me to set about the other two, but he asked what I required for what I had done " ‡. No man could engage in the practice of such a dangerous art in those days without making bitter enemies, and it is quite possible that this was one of the causes which conspired to develop in Wallis an overwhelming desire to rush into controversy where a wiser man would have hesitated.

But in spite of all this, his own and succeeding generations have every reason to be grateful to the memory of this versatile genius. Under his influence a brilliant mathematical school

* Stubbe, *A Severe Inquiry etc.*, p. 7.
† Wallis to John Fell, April 8, 1685 (Smith MS. 31, Bodleian).
‡ Add. MSS. 105, no. 128 (Bodleian) (June 6, 1702).

gradually arose. It may be said that in such an energetic age his mathematics soon became outworn. Certainly there is an archaic air about even his best work. But the same may be said about even Newton's, and, after all, its antiquity is a matter of style rather than outlook. Nearly three centuries of discoveries have been added since Wallis first directed his mind to the study of mathematics, but which of these discoveries cannot be traced in direct succession through Newton to Wallis ? For Wallis will always rank as Newton's great precursor.

His achievements were the outcome of persevering and unbroken study, mated to an extraordinary sagacity. " In the economy of her distribution ", says Brewster, " Nature is seldom lavish of her intellectual gifts ". Wallis lacked the powers of invention of Newton, and his position was won by sheer hard work. He was untiring, he was in his outlook original. His capacity for generalization has rarely been excelled, and it was by the judicious use of this that he prepared the way for the discoveries of the succeeding century. The principles of analogy and continuity which he developed, his interpretation of fractional and negative indices and the consequent widening of the horizon of mathematics were alone sufficient to ensure for him a distinguished place in the pages of the history of mathematics. His memory was prodigious, and never failed him almost to the closing years of his life. He often whiled away sleepless nights by exercises in mental arithmetic, and on one occasion he extracted the square root of a number of fifty-three digits, giving the answer correct to twenty-seven figures by an effort of memory alone.

Doctor Hearne wrote of him : " Doctor Wallis was a man of most admirable fine parts and great industry whereby in some years he became so noted for his profound skill in Mathematics that he was deservedly accounted the greatest person in that profession of any in his time. He was withal a good Divine, and no mean critic in the Greek and Latin tongues " *. Sir Peter Pett, in a letter to Pepys in 1696, spoke of him as " perhaps the greatest master of Algebra that ever lived " †. Even Aubrey wrote of him : " He hath writt severall treatises, and well; and to give him due prayse, hath exceedingly well deserved of the comonwealth of learning, perhaps no mathematicall writer so much " ‡.

Wallis had a real seventeenth century English hostility to the Church of Rome. We have already had an instance of this in his

* Smith MS. 54, no. 29 (Bodleian).
† *Correspondence of S. Pepys* (Tanner), p. 114. Sir Peter Pett to Mr. Pepys, May 3, 1696.
‡ Aubrey, II, ii, p. 570.

opposition to the Calendar Reform; the same attitude is to be noted in an unsigned fragment amongst his correspondence in the Bodleian. " Doctor Wallis ", says the writer, " I take to be a very worthy person, for whom I have a just esteem for the good service he did in the last reign in defence of the Protestant religion against Popery " *. Happily, his clashes with the Vatican were few, nor were they ever characterized by that bitterness which was so marked a feature of his other quarrels. Yet though the occasions upon which he crossed swords with Rome were not frequent, it is by no means unlikely that his attitude towards the Papacy had some bearing upon his deep-rooted antagonism towards France. For France, still dominated by the majestic figure of Louis XIV, was at this time on intimate terms with the Vatican, and it is quite possible that this was for Wallis a sufficient reason for regarding that nation with something more than suspicion.

From the letters which passed between him and his family we gather that Wallis was a kind and affectionate father. Moreover, though he had many enemies, he certainly had many friends, amongst whom Lord Brouncker and Sir Christopher Wren were conspicuous. Even Pepys, whose *Diary* records the fact " I met Wallis . . . . but he promises little " †, soon became warmly attached to him, and when Wallis sent him a copy of his works this evoked an affectionate reply from the great diarist ‡. Pepys' esteem and veneration for the now aged mathematician eventually led him towards immortalizing the person (for his name can never *dye*) " of that great man, and my most honoured friend " §. It was through his instrumentality that the magnificent portrait, which now hangs on the staircase of the Bodleian and which is reproduced in the Frontispiece, was painted. This portrait, which was executed by Kneller, bears the inscription :

Hanc
Magni WALLISII Oxoniensis
Effigiem
Celeberrimæ ACADEMIÆ Oxoniensi
D. D. D.
SAMUEL PEPYS
Car : & Jac : Angl : Regib : a Secretis
Admiraliæ.

* Add. MSS. 105, Fol. 119, June 15, 1700.
† Pepys' *Diary*, ii, p. 465, 16 Dec. 1666.
‡ *Correspondence of S. Pepys* (Tanner), vol. I, pp. xiii–xiv.
§ *Ibid.*, I, p. xiv.

APPENDIX I.

Brief Biographies of Wallis's Contemporaries and Immediate Predecessors, who are mentioned in this Work.

It is assumed that the works of Boyle, Descartes, Fermat, Galilei, Leibniz, Newton, Pascal are sufficiently familiar to the reader as to render their inclusion in the following pages unnecessary.

ANDERSON, ALEXANDER (1582–1619).—A native of Aberdeen, was a mathematician of note. He taught mathematics at Paris from 1612 to 1619, and whilst he was there, he became intimate with Vieta, whose works he edited during the years 1615 to 1617. The most important of these was the *De Æquationum Recognitione et Emendatione Tractatus duo*, which was published at Paris in 1615. This treatise contains Vieta's contributions to the solution of equations, together with an Appendix by Anderson himself, in which he shows that the solution of a cubic equation can be made to depend upon the trisection of an angle. He also published :—

(1) *Supplementum Apollonii Redivivi*. Paris, 1612. In this he shows a remarkable command over the ancient methods of Analysis.

(2) *Alexandri Andersoni Scoti ; Exercitationum Mathematicarum Decas Prima*. Paris, 1619.

AUBREY (AWBREY), JOHN (1626–97).—This famous antiquary was educated at Trinity College, Oxford. He was elected Fellow of the Royal Society in 1663 ; three years later he made the acquaintance of Anthony à Wood (*q. v.*), to whom he rendered much assistance in the publication of his *Antiquities of Oxford* (1674). Aubrey's correspondence with Wood continued until 1680, and his *Minutes of Lives* provided the latter with much of the material for his *Athenæ Oxonienses* (1690). Aubrey was a man of insatiable curiosity to which he combined a wide range of intellectual interests. He occupied himself for many years with a *History of Wiltshire*, and in 1695 he imparted his papers to Thomas Tanner (*q. v.*), Bishop of St. Asaph. The following year there appeared his *Miscellanies*, a highly entertaining collection of ghost stories and other anecdotes dealing with the supernatural. He also compiled a life of Hobbes, with whom he appeared to be on intimate terms, and whose philosophy he espoused. Although he recognised the great mathematical ability of Wallis, Aubrey does not appear to have entertained a high regard for his personal character. He died in 1697 at Oxford, and left much antiquarian and historical material in manuscript. According to the Dictionary of National Biography, his character as an antiquary has been unworthily traduced by Anthony à Wood, but has been fully vindicated by recent biographers. The MS. of his *Monumenta Britannica*, which was said to have been written at the command of Charles II, is in the Bodleian. The *Lives of Eminent Men*, by John Aubrey, Esq., was published in 1813.

APPENDIX I

BAKER, THOMAS (1625–90).—A mathematician of some eminence. He spent seven years at Magdalen Hall, Oxford, and was afterwards appointed Vicar of a parish in Devonshire. Here he devoted himself largely to mathematical studies. The result of his preoccupations was his treatise *The Geometrical Key, or the Gate of Equations unlocked : or a New Discovery of the Construction of All Equations however affected, not exceeding the Fourth Degree . . . . and the Finding of all their Roots, as well False as True*, etc. (Lond., 1684). This contains a general account of the solution of biquadratic equations by means of a parabola and a circle, described by Montucla as " une invention fort élegante et ingénieuse ".

The work is reviewed in the *Philosophical Transactions* for 1684 (vol. xiv, no. 157, p. 549). The Royal Society sent Baker a number of mathematical queries, which he answered so satisfactorily that the Society presented him with a medal bearing an honourable inscription.

BALIANI, GIOVANNI BATTISTA (1582 or 6–1660 or 6). — Captain of Archers, was an amateur mathematician. His best known work was his *De Motu Naturali Gravium Solidorum et Liquidorum* (Genoa), which first appeared in 1638. In this edition he enunciated the " law " that the velocity acquired by a falling body must be proportional to the space already described, and for this, the work earned the condemnation of Montucla. In a later edition of his work (Genoa, 1646), Baliani enunciated the correct law, namely, *Gravia naturali motu descendunt semper velocius ea ratione, ut temporibus æqualibus descendant per spatia semper majora, juxta proportionem quam habent impares numeri ab unitate inter se* (Prop. V, p. 25), *i. e.*, the spaces described by a falling body in successive seconds of its motion are proportional to the *odd* numbers. He also published *Opere Diverse* (Genoa, 1666), in which he solved various problems in mechanics.

BARROW, ISAAC (1630–77).—This eminent divine and mathematician went to school, like Wallis, at Felsted, whence he proceeded to Trinity College, Cambridge. He rapidly distinguished himself as a scholar in many branches of learning, and in 1649 he was elected Fellow of his College. He left Cambridge soon afterwards, however, and travelled throughout Europe, visiting Constantinople. On his return in 1660, he was ordained, and the same year he was chosen Greek Professor at Cambridge. In 1662, on the death of Lawrence Rooke, he was chosen Gresham Professor of Geometry, and the following year he was elected Fellow of the Royal Society. About the same time he was named the first Lucasian Professor of Mathematics, but soon afterwards he resigned this Chair to Newton, whose superior abilities he recognised. He also successfully cultivated the science of Optics. His *Lectiones Opticæ* is described by Montucla (ii, 505) as " une mine de propositions optiques, curieuses, intéressantes, et auxquelles la géométrie est toujours appliquée avec une élégance particulière", and James Gregory was so impressed with it that he wrote to Collins : " Mr. Barrow in his

Optics, sheweth himself a most subtle geometer, so that I think him superior to any that ever I looked upon " *. In 1670 he was created D.D. ; in 1672 he became Master of Trinity, and three years later, Vice-Chancellor of the University. The last few years of his life were devoted entirely to Divinity.

Like Wallis, Barrow will always rank as one of the great precursors of Newton. His mathematical works were very numerous. These had a singular conciseness which, however, did not destroy their clarity. The chief of them were :—

(1) *Euclidis Elementa.* Camb., 1655 (Latin) and 1660 (English).

(2) *Euclidis Data.* Camb., 1657.

> In both these he follows the notation adopted by Oughtred.

(3) *Lectiones Opticæ*, xviii. Lond., 1669.

(4) *Lectiones Geometricæ*, xiii, Lond., 1670. This contains his method of drawing a tangent at a given point on a curve by means of his " differential triangle", and is replete with profound researches on the properties of curved figures and tangents to them.

(5) *Archimedis Opera, Apollonii Conicorum Libri IV* (1675).

> His *Mathematicæ Lectiones*, xxiii (1664–1666) were published in 1685. His complete works were published under the title of " The Works of the Learned Isaac Barrow, D.D." in four volumes in 1683–7 by the Rev. Dr. Tillotson, and by Whewell in 1880. Like Oughtred, Barrow was devoted to the Royalist cause.

BARTHOLINUS, ERASMUS (1625–98).—An eminent Danish mathematician who successively became Professor of Mathematics and of Philosophy at Copenhagen. He was one of the chief promoters of the new geometry throughout Denmark. Whilst travelling through France, he became acquainted with De Beaune, from whom he obtained much valuable material in manuscript. This he communicated to Van Schooten, who used it to enrich the second edition of his Commentary on Descartes. His lesser known works include *Dimensiones Æquationum*, 1664 ; *De Problematibus Geometricis* ; *per Algebram solvendis Dissertationes*, viii, 1664. His major works were printed in Copenhagen, and include :—

(1) *De Cometis Anni* 1664 5.

(2) *Opusculum* : *Ex Observationibus Hafniæ habitis adornatum.* Hafniæ, 1665.

(3) *Experimenta Crystalli Islandici.* Hafniæ, 1669.

(4) *Selecta Geometica.* 1674.

* Gregory to Collins, 29 Jan. 1670. Rigaud, ii, 190.

APPENDIX I

BENEDETTI, GIOVANNI BATTISTA (1530-90).—A writer on mathematical subjects. In his *Diversarum Speculationum Mathematicarum et Physicarum Liber* (Taurini, 1585), a work too little known, he explained the acceleration of falling bodies as due to a *virtus impressa*. He held the correct view regarding centrifugal force, for, contrary to the Aristotelian doctrines, he maintained that a body moving under constraint would when released immediately describe a straight line. He was intimate with Galilei, whose work he probably influenced. He also wrote *De Gnomonum Umbrarumque Solarium Usu* (Taurini, 1574).

BOMBELLI, RAFFAELLO (16th century).—Of Bologna; published an Algebra in 1572 at Venice, and again in 1579 at Bologna. This existed in manuscript at least twenty years before it appeared in print. In this work Bombelli demonstrated the reality of the roots of the cubic in the so-called irreducible case. His views on Imaginaries were well in advance of his time. Following Cardan he solves the equation $x^3 - 7x = 6$, and he finds that x is equal to

$$\sqrt[3]{3 + \sqrt{\frac{-100}{27}}} + \sqrt[3]{3 - \sqrt{\frac{-100}{27}}}.$$

which eventually reduces to $x = 3$.

He uses p and m for plus and minus, and he indicates cube roots thus, ℞c. He also introduced the signs ⌊ ⌋ for the root of a compound expression, instead of the sign ℞V (*Radix Universalis*).

Wallis attributes to Harriot many of the discoveries which are to be found in Bombelli; further he affirms that Descartes made free use of Bombelli's methods. Both these statements are sharply criticised by Montucla. After having accused Wallis of having read Bombelli " avec beaucoup d'inattention ", he observes that " il falloit être aveuglé comme l'étoit Wallis, par l'envie de déprimer le géomètre François, pour tomber dans une pareille inexactitude ", and he adds that Wallis would not have noticed the priority of Bombelli's work over Descartes' had Harriot been in the place of Descartes. (Montucla, i, p. 598.)

BORELLI, GIOVANNI ALFONSO (1608-79).—A distinguished Italian physicist and physiologist. He became Professor of Mathematics at Messina and at Pisa. In this capacity he maintained an active correspondence with many foreign scholars, including Collins. His best known work is his *De Motu Animalium* (Rome, 1680-5). In this he sought to explain the movements of the animal body on mechanical principles. In addition, he published several mathematical treatises, the chief of which were :—

(1) *De Vi Percussionis*. Bologna, 1667.

(2) *De Motionibus Naturalibus a Gravitate Pendentibus*. Reg., 1670.

(3) *Elementa Conica Apollonii Pergæi et Archimedis Opera. Nova et breviori Methodo Demonstrata.* Rome, 1679.

Borelli was elected to the Accademia del Cimento in 1657, and eventually he rose to be one of its most illustrious members.

BRIGGS, HENRY (1556–1631).—A mathematician of repute. He was educated at St. John's College, Cambridge, where in 1588 he was elected Fellow. He became the first Professor of Geometry at Gresham College in 1596, and in 1619 he was appointed to the Savilian Professorship of Geometry in succession to Sir Henry Savile himself. This latter post he held until his death in 1631. In 1617 he constructed the first Table of Common Logarithms from 1 to 20,000 and from 90,000 to 100,000, which rapidly became popular throughout western Europe. These Tables were prefaced by an erudite introduction, in which the theory as well as the use of these numbers was clearly indicated, and in which the origin of the method of interpolation was demonstrated. Briggs was largely responsible for the introduction of the decimal notation for fractions. His notation was an improvement upon that of Stevinus, for he underlined the decimal part. Thus he wrote the decimal 34·651 as 34651. His best known works were :—

(1) *A Description of an Instrumental Table* (1616).

(2) *Euclidis Elementorum. Sex libri priores.* Lond., 1620.

(3) *Joh. Neperi Arithmetica Logarithmica.* Lond. (1624).

(4) *Trigonometria Britannica. Libri duo.* (1633).

(5) *Tables for the Improvement of Navigation* (1610).

In addition, Briggs was no mean scholar of astronomy.

BROUNCKER, WILLIAM, Lord Viscount of Castle Lyons (1620–84).— Though apparently designed for the study of medicine (he became M.D. at Oxford in 1647), he early manifested a genius for mathematical studies, and it is by his mathematical writings that he is best remembered. He was one of the original members of the Royal Society, and after its Incorporation he became its first President, a post which he held for fifteen years. He was also President of Gresham College from 1664 to 1667, and in 1662 he became Chancellor to Queen Catharine.

By writing Wallis's famous π series in the form given in Chapter IV, Brouncker first rendered familiar the theory of continued fractions. Following the methods elaborated in the *Arithmetica Infinitorum*, he also gave a series for the quadrature of the equilateral hyperbola. This was mentioned by Wallis in a tract against Meibomius in 1657, though Brouncker delayed publication of it until it appeared in the *Philosophical Transactions* for 1668.

During the years 1657-1658, he engaged in a correspondence on mathematical subjects with the French mathematicians. This was afterwards published under the title *Commercium Epistolicum.*

APPENDIX I

Brouncker wrote several learned papers on mathematical subjects. Chief among these were :—

(1) *Experiments on the Recoiling of Guns.* (Sprat, p. 233.)
(2) *The Squaring of the Hyperbola by an Infinite Series of Rational Numbers together with its Demonstration.* (*Phil. Trans.*, 1668.)
(3) *On the Proportion of a Curved Line of a Paraboloid to a Straight Line, and of Finding a Straight Line equal to that of a Cycloid.* (*Ibid.*)

In addition, there are several letters to Wallis on mathematical subjects.

BYFIELD, ADONIRAM (died 1660).—Was Wallis's colleague as Secretary to the Westminster Assembly of Divines. Like Wallis he was educated at Emmanuel College, Cambridge. A strong Puritan, he became Chaplain to the Parliamentary Forces. Afterwards he became Vicar of Fulham. In 1654 he was nominated one of the Assistant Commissioners for Wiltshire, under an Ordinance dated June 29th, for ejecting "Scandalous, ignorant, insufficient Schoolmasters and Ministers"*.

CARCAVI, PIERRE DE (date of birth not known).—Was born at Lyons, and became Counsellor to the Parliament of Toulouse after Fermat. In 1666 he was appointed geometrician to the Académie des Sciences, of which he was a Charter Member. He maintained a long correspondence with Mersenne, and there are several important letters of his amongst those of Descartes. In 1645 he demonstrated the impossibility of the quadrature of the circle. He died in Paris in 1684.

CARDAN, GIROLAMO (1501–75 or 76).—Appears to have dabbled in most branches of science, particularly medicine, before turning his attention to mathematics. He studied at Pavia and at Padua, and became Professor of Mathematics at Milan in 1533. He was one of the greatest algebraists of his time, and in his *Ars Magna* (1539) he illustrated the solution of cubic equations by the so-called Cardan's rules†. He does not omit to take notice of negative roots (which he calls *feintes*), for in Art. 7 of the *Ars Magna* he proposes the equation $x^2+4x=21$, and he remarks that the roots are $+3$ and -7. In certain cubic equations he discovers all three roots. In his earlier works he rejects imaginary roots, but in the treatise mentioned (*Quæstio* xvi) he solves the problem : Divide 10 into two parts whose product is 40, and he gives as his solution $5+\sqrt{-15}$ and $5-\sqrt{-15}$, which he writes thus :

$$5p : \mathrm{R}m : 15 \quad \text{and} \quad 5m : \mathrm{R}m : 15.$$

---
* *Dict. Nat. Biog.*, viii, p. 111.
† To solve an equation of the type $x^3+px=q$ he puts $ab=\frac{1}{3}p$, and $a^3-b^3=q$. These two equations lead to a quadratic in a^3, from which the values of a and b may be obtained. Since $x=a-b$, this leads to the determination of one root of the given equation.

He also tried to apply geometry to physics in a work entitled *Opus novum de Proportionibus, Numerorum, Motuum, Ponderum, Sonorum, etc.* (Basil, 1570). All his mathematical works are to be found in the *Opera Cardani* (Lyons, 6 vols.). Montucla says of him : " Il est vrai que ses efforts furent en général destitués de succès ; les bases propres à fonder ses raisonnemens géométriques manquoient ; mais cet ouvrage qui lui fit beaucoup d'honneur, montre qu'il avoit dans la tête autant de géométrie qu'aucun de ses contemporains " (i, 571).

CASTELLI, BENEDETTO (1577–1644).—A friend of Galilei. He is best known by his treatise on hydrostatics, which bore the title *Treatise on the Mensuration of Running Waters* (Della Misura dell' Acque Correnti) (Rome, 1628), which was said to have been written at the request of Pope Urban VIII, who entertained fears of disaster due to the river overflowing its banks. This treatise appeared in 1638. It was translated into French in 1644, and became popular throughout Europe. It was described by Montucla as " un ouvrage peu considérable par le volume mais précieux par la solide et judicieuse doctrine qu'il contient ".

CASWELL, JOHN (fl. 1680–1700).—Was M.A. of Wadham College, Oxford, and afterwards Vice-Principal of Hart Hall. He observed the height of the barometer on Snowdon in 1686, and he obtained the quadrature of a portion of the epicycloid. (*Phil. Trans.*, 1695). He wrote a work on Trigonometry, which was published with Wallis's *Algebra*. In this he introduced a number of symbols. He wrote S for *sine*, Σ for *cosine*, and T for *tangent*.

CAVALIERI, BONAVENTURA (1598–1647).—This celebrated mathematician was born at Milan in 1598. In his early days he showed such zeal for his studies that he was sent to Pisa. There he met Castelli, who stimulated his interest in mathematics. He eventually became Professor of Mathematics at Bologna, and his papers show that he was in possession of the Method of Indivisibles as early as 1629. This, however, was not published until 1635, when it appeared under the title *Geometria Indivisibilibus Continuorum nova quadam Ratione Promota* (Bologna). This treatise, which was published again in 1653, revived interest in the subject, and encouraged Wallis to prosecute it still further. Cavalieri also published a treatise on Conic Sections with the title *La Specchio Usterio, ovvero Trattato delle Settione Coniche* (Bologna, 1632), and a system of Trigonometry with the title *Directorium Generale Uranometricum* (1632). His *Exercitationes Geometricæ Sex*, which appeared in 1647, contains examples on the method of Indivisibles. In one of these he showed that his method was no other than the Method of Exhaustions which had been employed by the ancient geometers.

APPENDIX I

CLAVIUS, CHRISTOPHER (1537–1612).—A studious mathematician, who eventually became Professor of Mathematics at Rome. He wrote an Algebra which was published at Rome in 1608. This is a somewhat elementary treatise, though it has the credit of having introduced the German algebraic symbolism into Italy. For the most part he follows Stifel, and in his treatment of equations he does not touch anything above the quadratic. He also wrote *Epitome Arithmeticæ Practicæ* (Rome, 1583). We have from his pen also a new edition and translation of Euclid (1574), which, though somewhat prolix, is, nevertheless, a very able commentary. Clavius was commissioned by Pope Gregory XIII with the task of assisting in the exposition of the Gregorian Calendar. His mathematical works were published in 1612 in five volumes under the title *Opera Mathematica* (Moguntiæ, 1611–12).

COLLINS, JOHN (1625–83).—Was the friend and counsellor of many of the most distinguished scientists of his day. He was a man of sound judgment and ability and, though of humble origin, he rapidly acquired a commanding position in science, becoming a Fellow of the Royal Society in 1667.

Throughout his life he maintained an intimate correspondence with many of the most learned men at home and abroad. Indisputably his greatest service to science was the fact that he promoted the publication of many valuable works which might otherwise have been lost to the world. Chief among these are Barrow's *Optics* and his *Geometrical Lectures* ; Branker's *Translation of Rhonius's Algebra, with Doctor Pell's Additions*. Amongst his papers are MSS. on mathematical subjects by Briggs, Oughtred, Newton, Barrow, and Pell, and from these it is quite clear that he spared neither pains nor expense to promote the advancement of science. He seems to have been the clearing house of all the improvements made by scholars of that period, so much so that he acquired the title of the English Mersenne. Aubrey, the antiquary, " ownes the progress of mathematical learning overmuch to his industry therein " *.

The *Commercium Epistolicum, D. Johannis Collins, et aliorum De Analysi Promota ; jussu Societatis Regiæ In Lucem editum* (1712), which gives priority of the invention of the Calculus to Newton, was composed from letters in his possession. In addition, he published a number of pamphlets on Trade and on Navigation, e.g., *The Description and Uses of a General Quadrant* (Lond., 1658). From a letter to Wallis, dated October 3, 1682, we gather that Collins designed the publication of a *Treatise on Algebra*. The completion of this work was, however, forestalled by his death.

COTES, ROGER (1682–1716).—This brilliant mathematician was educated at St. Paul's School, and at a very early age he manifested a genius for mathematical studies. He became a Fellow of Trinity College in 1705, and Plumian Professor of Astronomy the year following. He became F.R.S. in 1711. He helped Newton to re-issue the *Principia*

* Aubrey MSS. 13, fol. 343, Sept. 17. 1683 (Bodleian).

in 1713, and this greatly enhanced his reputation. He published *Logometria* ; *A Treatise on Ratios* (*Phil. Trans.*, xxix, 1714), and he made the earliest attempt to frame a theory of Errors. He also wrote a *Compendium of Arithmetic* ; *On the Resolution of Equations* ; *of Dioptrics* ; and *On the Nature of Curves*. Newton had a profound admiration for his abilities, and he wrote of him : " If Cotes had lived, we might have known something ". His mathematical papers were published in 1722 and in 1738 by Dr. Smith, his successor, and his correspondence in 1850.

CURLL, WALTER (1575–1647).—A noted Divine, who was educated at Peterhouse College, Cambridge. In 1612 he became Fellow of his college, and in the same year he was made D.D. He became Dean of Lichfield in 1621, Bishop of Bath and Wells (1629), and ultimately Bishop of Winchester (1632). He helped to defend Winchester Castle against the hordes of Cromwell in 1645. He was compelled to surrender, and was deprived not only of his episcopal income, but even of his private property.

DIGBY, SIR KENELM (1603–65).—Author, naval commander, and diplomatist. He entered Gloucester Hall (Worcester College), Oxford, in 1618. Two years later he visited Paris. In 1649 he was banished and went to France, where he made the acquaintance of Descartes. He returned to England in 1654, and eventually he became a member of the Council of the Royal Society. He was a scientific amateur, rather than a man of science. He believed in astrology and in alchemy, and he wrote *A Choice Collection of Rare Chymical Secrets* and *A Treatise on the Nature of Bodies* (Paris, 1644). Wallis, Wilkins and Ward all speak admiringly of him, and the last named addressed him as the " ornament of this nation ".

DU LAURENS, FRANÇOIS.—A slight French mathematician, who flourished during the latter half of the seventeenth century. He wrote *Specimina Mathematica duobus libris Comprehensa* (Paris, 1667), a copy of which is in the British Museum (8532 dd 6), but as none of the usual French reference books even mention his name, he could not have achieved much success as a writer on mathematics. He used a number of symbols, such as the dot . for ratio, and :: for proportion, ⊓ for equal to, ⌐ for greater than, ⌐ for less than, $\boxed{3}$ for cube, etc. His chief title to fame appears to be that the work mentioned above was one which roused Wallis to pour contempt upon its author, who, according to Wallis, was " much more negligent of what he writes than doth become a mathematician "*. " As for Du Laurens ", he wrote to Collins, " I do not look upon him to have any great matter, but what he hath from others, and what notions he hath were but crude and undigested, and of which he was not at all master " †. Many of his letters to Oldenburg are in the same strain, and in one of them he declared that Du Laurens' work would have been better styled *Negligentiæ Specimina*.

* To Oldenburg, Mar. 30, 1668. (*Royal Society Guard Books.*)
† Rigaud, ii, 559–60.

APPENDIX I

ENT, SIR GEORGE (1604–89).—A noted physician, was the son of Josias Ent, a merchant of the Low Countries who had been driven by religious persecution to England. He studied at Sidney Sussex College, Cambridge, where he graduated M.A. in 1631. He afterwards became M.D. at Padua in 1636. He was appointed Gulstonian Lecturer in 1642, and was knighted in 1655. He was President of the Royal College of Physicians from 1670 to 1675 and in 1682 and 1684. He was one of the group whose enthusiasm led to the formation of the Royal Society, and he became one of its Charter Members. His chief work was a tract, *Apologia pro Circuitione Sanguinis* (1641), in which he vindicated the discovery of Harvey, whose friend he was.

FABRI, HONORÉ (1606–88).—A noted French Jesuit, theologian, physicist, and astronomer. He was a distinguished member of the Accademia del Cimento. About 1660 he wrote a treatise on the Cycloid (*Opusculum Geometricum de Linea Sinuum et Cycloide*), and it is said that he might have attained a high rank in the field of science, had he not been distracted by other duties. His published works were:

(1) *Synopsis Optica*, 1667. (This was reviewed by Wallis in a letter to Oldenburg, March 7, 1667/8.)

(2) *Dialogi Physici*, 1669.

(3) *Synopsis Geometrica*, 1669.

(4) *Physica seu Scientia Rerum Corporearum*, 1669–71.

In the dispute between Fabri and Borelli, Wallis took the side of Borelli. " I know not what character to give him (Fabri) ", he wrote to Oldenburg, " For to declare for him will not be fit unless there were more reasons. And to declare against him would sett him a-wrangling, being (if I mistake him not by his writings) a conceited quarrelsome man " *.

FELL, JOHN(1625–86).—Was educated at Christ Church, Oxford, of which his father Samuel Fell was Dean. On the outbreak of the Civil War he took up arms for the King, and was in consequence ejected from his studentship by the Parliamentary Visitors in 1648. " He was the most zealous man of his time for the Church of England ", says Wood, " and none that I yet know of did go beyond him in the performance of the rules belonging thereto ". Eventually he became Dean of Christ Church, and Bishop of Oxford.

He also became Vice-Chancellor of Oxford in 1666, and he displayed great zeal in maintaining a high standard of discipline, and he was even instrumental in expelling his friend Locke.

In addition to other works he published *Grammatica Rationis sive Institutiones Logicæ* (Oxon, 1673 and 1685) and *Historia et Antiquitates Universitatis Oxoniensis*, 1674.

* To Oldenburg, Dec. 12, 1676. (*Royal Society Guard Books*.)

FLAMSTEED, JOHN (1646–1719).—The first Astronomer Royal, developed an interest in astronomy when quite young. In 1669, he calculated some remarkable eclipses of the moon and sent them to Lord Brouncker. These were greatly applauded by the Society, and brought him to the notice of many distinguished men of science including Collins and Oldenburg. Whilst at Cambridge he became acquainted with Newton, Barrow, and other learned scientists, and in 1677 he was elected to a Fellowship of the Royal Society.

Though his health was far from robust, Flamsteed was a man of amazing energy. Between 1676 and 1689 he made observations on no less than 20,000 star places. He was keenly jealous of his professional reputation, and as a result he drifted into a position of antagonism with many of his scientific contemporaries. His observations were of great service to Newton in the production of the *Principia*. In 1707 the first volume of his *Catalogue of Stars* appeared, but disputes arose with Newton and Halley, who published without Flamsteed's consent an imperfect edition of his latest observations. For this he called Halley a " lazy and malicious thief " (*D. N. B.*, xix. 246).

Flamsteed's great work, *Historia Cœlestis Britannica* (3 vols.), was published in 1725 by his assistant, Joseph Crosthwait. Astronomy is also indebted to him for two excellent treatises, *De Temporis Æquatione* (Lond., 1673) and the *Lunar Theory of Horrocks* (Lond., 1679). In addition he has many papers in the *Philosophical Transactions*. Flamsteed showed himself equally skilled in the practical parts as well as the theoretical parts of astronomy by ascertaining absolute Right Ascension through simultaneous observations of the sun and a star near both equinoxes, which is the basis of modern astronomy.

FOSTER, SAMUEL (?–1652).—Was one of the group whose zeal led to the birth of the Royal Society, though he did not live to see its Incorporation. Like Wallis, he received his education at Emmanuel College, Cambridge, where he displayed marked enthusiasm for astronomical studies. He became Gresham Professor of Astronomy in 1636, and again from 1641 to 1652, having been temporarily expelled for refusing to kneel at the Communion Table. He published many astronomical works, including *The Description and Uses of a Small Portable Quadrant* (Lond., 1624 and 1652), *The Art of Dialling* (Lond., 1638 and 1675). Other works of his were published posthumously. Dr. John Twysden gives him the character of " a learned, industrious, and most skilful mathematician " *.

FRÉNICLE DE BESSY, BERNARD (1605–75).—This distinguished French mathematician was a contemporary and an intimate friend of Fermat and Descartes. In 1666 he was admitted geometrician to the French Académie des Sciences. He wrote many papers on mathematical subjects, which were published in the Memoirs of the Académie,

---

* Preface to Foster's Miscellanies. (*Miscellanies : or Mathematical Lucubrations of Mr. Samuel Foster sometime Publike Professor of Astronomie in Gresham Colledge in London*, by John Twysden (Lond., 1659).)

APPENDIX I

particularly Volume V. These relate mainly to the Theory of Numbers, Combination and Magic Squares. His correspondence with Wallis has been described in Chapter V. " Frénicle I take to be very good at numeral questions ", wrote Wallis, " such as those of Diophantus, to which he had a peculiar genius " (Rigaud, ii, 560). In a letter to Wallis, dated Dec. 20, 1661 (Huygens, *Œuvres*, Tom. iv, p. 45), he uses = for equal.

GASSENDI, PIERRE (1592–1655).—A celebrated French philosopher, who rose to be Professor of Philosophy at Aix before he was twenty. He developed a great aptitude for astronomy, and was the first person to observe the transit of Mercury across the sun (Nov. 7, 1621). In 1645 he was appointed King's Professor of Mathematics at Paris, an institution intended chiefly for astronomy. From his letters it appears that he was often consulted by some of the most celebrated men of his time, including Kepler, Hevelius and Galilei. He wrote :—

(1) Three volumes of Epicurus's *Philosophy*.

(2) *Astronomical Works*.

(3) *Lives of Epicurus, Copernicus, Tycho Brahe, Purbach and Regiomontanus*. Paris, 1654.

(4) Epistles, and other treatises.

All his works were collected together and printed at Lyons in 1656 (*Opera omnia*, 6 vols. folio, 1658). He wrote much against the philosophy of Descartes.

GIRARD, ALBERT (1595–1634).—Was an ingenious Flemish or Dutch mathematician. He published an edition of Stevin's *Arithmetic* in 1625, augmented with many notes. His chief work, however, was his *Invention Nouvelle en l'Algebra, tant pour la Solution des Equations que pour recognoistre le Nombre des Solutions qu'elles reçoivent; avec plusieurs Choses qui sont necessaires a la Perfection de ceste divine Science* (Amsterdam, 1629.) Following Stevin he denotes powers of his unknown by symbols ⓪, ①, ②, etc., and their roots by $(\frac{1}{2})$, $(\frac{1}{4})$, and sometimes by the more usual $\sqrt[2]{}$, $\sqrt[3]{}$, $\sqrt[4]{}$. He solves quadratics by completing the square, taking both positive and negative roots, and he observes that sometimes the equation is *impossible* as

② equ. 6 ① —25, (i. e., $x^2 = 6x - 25$),

whose roots he adds are $3 + \sqrt{-16}$ and $3 - \sqrt{-16}$.

He solves cubics mainly by Cardan's rules. In this treatise he shows that in the cubic equation which leads to the irreducible case, there are always three roots, two positive and one negative, or conversely. Here we find a knowledge of negative roots (which he calls " *par moins* ") more completely developed than in almost any other analyst. Another remarkable feature of this treatise was that in it Girard showed the use of negative roots in geometry eight years before Descartes, by these words : " La solution par moins s'explique en géométrie en rétrogradant, et le moins recule où le $+$ avance " ; and he illustrates this by a problem which leads to an equation of the fourth degree, where two roots are positive and two are negative. He infers that every algebraical equation admits of as many roots as there are units in the index of the highest power, and these roots may be either positive, negative or imaginary, or as he calls them, greater than nothing, less than nothing, or involved. Thus the roots of the equation

$$1\,\text{\textcircled{3}}\, \text{equ.}\,7\,\text{\textcircled{1}}-6,\ \text{i. e.,}\ x^3=7x-6,$$

are 2, 1, and -3, and the roots of the equation

$$1\,\text{\textcircled{4}}\,\text{eq.}\,4\,\text{\textcircled{1}}-3,\ \text{i. e.,}\ x^4=4x-3,$$

are 1, 1, and $-1+\sqrt{-2}$ and $-1-\sqrt{-2}$.

Girard also wrote on Spherical Trigonometry.

GLISSON, FRANCIS (1597–1677).—A physician and a pioneer in physiology, was one of the most distinguished anatomists of his time. He graduated M.A. at Caius College, Cambridge, in 1634, and became Regius Professor of Physic in 1636, a post which he held until the year of his death. He was a great friend of Harvey, and he was also associated with those whose enthusiasm led to the formation of the Royal Society. He was President of the Royal College of Physicians from 1667 to 1669. He published numerous medical monographs, chief of which was his *Anatomia Hepatis* (1654). Wallis was one of his pupils. His medical works were collected and published in 1691 under the title : *Opera Medica Anatomica*.

GODDARD, JONATHAN (1617–75).—A distinguished physician, who graduated M.B. at Christ's Hospital, Cambridge, in 1638, M.D. in 1643, and in 1648 he became Gulstonian Lecturer. As Physician-in-Chief, he accompanied Cromwell to Ireland in 1649, and to Scotland in 1650. According to Wood, " by the said Oliver's powers, he became Warden of Merton College in 1651, and was afterwards elected Burgess to serve in the Little Parliament (1653) and was made one of the Council of State ". In 1655 he was made Gresham Professor of Physic, and later he was elected to a Fellowship of the Royal Society, and membership of its Council. Upon the Restoration, he was ejected from his Fellowship of Merton, and lived mostly at Gresham College, where he had a laboratory for the preparation of medicines and for the devising of experiments for the Royal Society. He was a particularly industrious member of the Society and, according to Aubrey, a " zealous member for the improvement of naturall knowledge among them. They made him their drudge, for when any curious experiment was to be donne, they would lay their taske on him " *.

He wrote two *Discourses against Apothecaries* and *Experiments on Refining Gold with Antimony* (*Phil. Trans.*, 1677). According to Seth Ward, he was the first Englishman to make a telescope.

GRÉGOIRE DE SAINT-VINCENT (1584–1687). *See* SAINT-VINCENT.

GREGORY, DAVID (1661–1708).—Was a man of acute and penetrating genius. He was educated at Aberdeen and later at Edinburgh. At the age of 23 he was elected Professor of Mathematics at Edinburgh, and in the same year he published his *Exercitatio Geometrica de Dimensione Figurarum* (Edinburgh, 1684). He very soon perceived the merits of the Newtonian philosophy, and to him is due the credit of introducing it into Edinburgh, whilst less fortunate scholars still laboured, according to Whiston, under the Cartesian philosophy.

In 1691 he was appointed Savilian Professor of Astronomy, largely through the instrumentality of Newton and Flamsteed. The appointment was strongly opposed by Hearne, who attacked his astronomy and accused him of having used material from the *Principia* without due acknowledgment, " which, as 'tis said has put Sir Isaac on a new edition of his *Principia* ". He was elected Fellow of the Royal Society in the following year.

The idea of achromatic lenses occurred to him in 1695. In 1702 he published *Astronomiæ Physicæ et Geometricæ Elementa*. This, which is founded upon the Newtonian doctrines, is generally considered his masterpiece. In addition, he has seven papers in the *Philosophical Transactions*.

GREGORY, JAMES (1638–75).—Uncle of the above, was an original discoverer in mathematics. He was educated at Aberdeen. He travelled abroad, and at the age of 24 he published his *Optica Promota* at Padua, in which his reflecting telescope was described. This discovery at once attracted the attention of mathematicians both in this country and abroad. He came to London in 1664, and became intimate with Collins. In 1668 he published his *Vera Circuli et Hyperbolæ Quadratura* at Patavia, in which he propounded his discovery of the infinitely converging series for the area of the circle and of the hyperbola. A copy of this was sent to the Royal Society, where it received the commendation of Lord Brouncker. This treatise, however, provoked a controversy with Huygens (*q. v.*), and with Wallis, the latter maintaining (in a letter to Brouncker, Nov. 4, 1668) that Gregory's proof that the circle could not be squared was not rigorous. Gregory became a Fellow of the Royal Society in 1668, and he engaged in a voluminous correspondence with some of the most illustrious mathematicians of the age, including Newton. In the same year he published his *Exercitationes Geometricæ* (Lond., 1668), which greatly enhanced his

* Aubrey, ii, p. 358.

reputation. Soon afterwards he became Professor of Mathematics at St. Andrews, and later at Edinburgh (1674). He also invented and demonstrated by the help of the hyperbola a very simple converging series for computing logarithms. In addition to the above, Gregory wrote *The Great and New Art of Weighing Vanity* (1672) (an attack upon Sinclair, ex-Professor of Philosophy at Glasgow), as well as several important papers which were published in the *Philosophical Transactions* and in the *Commercium Epistolicum* of John Collins. His death, at the early age of 37, probably robbed him of the title of Newton's greatest contemporary.

HAAK, THEODORE (1605–90).—Was born at Neuhausen, and came to England in 1625. He translated many English works into High Dutch. He is of importance mainly on account of the part he played in the formation of the Royal Society. He did not contribute anything of his own to the *Transactions*, but he communicated Hooke's *Philosophical Collections* for February, 1681–2, as well as Mersenne's and Descartes' criticisms of Pell's *An Idea of the Mathematics*. During the Civil War, he espoused the cause of the Parliament. His portrait hangs in the Bodleian.

HALLEY, EDMOND (1656–1742).—This celebrated astronomer was educated at St. Paul's School, London, and at Queen's College, Oxford. As a boy he gave evidence of ardour for learning. His reputation became firmly established when in 1677 Charles II sent him to make observations of the stars in the Southern Hemisphere. He resided at St. Helena for two years, and whilst he was there, he laid the foundations of Southern Astronomy. On his return, he compiled with amazing care and accuracy his *Catalogus Stellarum Australium* (1678), and shortly afterwards he was elected a Fellow of the Royal Society. In 1678 he arbitrated between Hooke and Hevelius, and in 1680 he made cometary observations with which his name is associated. He became Clerk to the Royal Society, and he edited its *Transactions* from 1685 to 1693. He was largely instrumental in seeing the *Principia* through the press.

Whilst in command of the *Paramour Pink*, he explored the Atlantic and prepared a General Chart of the Variation of the Compass, with the Halleyian Lines, 1699–1702. On the death of Ward, he was nominated to the vacant Savilian Professorship of Astronomy. He was not appointed, on account of the suspicion, which he vainly tried to combat, of his holding materialistic views. When however, the Savilian Chair of Geometry became vacant, on the death of John Wallis, Halley was nominated to fill it, and from that year until his death he devoted himself chiefly to geometry. He prepared Flamsteed's observations for the press, and he edited the first version (1712) of Flamsteed's *Historia Cœlestis*.

Halley became Secretary of the Royal Society in 1713, and Astronomer Royal in 1721 on the death of Flamsteed. He was also a foreign

APPENDIX I

member of the Académie des Sciences. By his Breslau *Tables of Mortality* he may be said to have originated the science of life statistics. Besides a multitude of contributions to the *Transactions* on a variety of subjects, we have from him :—

Astronomiæ Cometicæ Synopsis, Oxf., 1705.

Apollonii de Sectione Rationis, Libri duo, 1706.

Apollonii Pergæi Conicorum, Libri viii, 1710.

HARRIOT, THOMAS (1560–1681).—Was an eminent English algebraist, and quite possibly an astronomer of note. After graduating M.A. at St. Mary's Hall, Oxford, he accompanied Raleigh on his expedition to Virginia. On his return he gave himself up to mathematical studies, particularly algebraical analysis, and to him is due the credit of having made the first advances in the subject since Vieta. He made some important discoveries regarding the nature and formation of equations, the outlines of which had quite possibly been observed by Vieta, but which were developed with much greater fullness by Harriot. It was he who first decomposed equations into their simple factors, but as he did not recognize negative or imaginary roots, he failed to prove that every equation could be so decomposed. His work in this branch of mathematics was expounded in his *Artis Analyticæ Praxis* (1631). On account of the frequent references to this treatise in the preceding pages, it is necessary to give a more detailed account of it here. There is a Foreword in which the contributions of Stevinus and Vieta are eulogised. This appears to have been written by Walter Warner the publisher.

The *Praxis* is prefaced by eighteen Definitions, chiefly on Canonical Equations. There are six sections, which may be briefly summarized thus :—

Sectio I. Logistices Speciosæ Quatuor operationum formæ exemplificatæ. In this are explained the fundamental operations of arithmetic. He also shows how to arrange an equation in a form convenient for solving (*Æquationum irregularium ad formam legitimam reductiones exemplificatæ*), e. g., in descending powers of the unknown ; the first term is made positive, and its coefficient unity. In this section (p. 10) he uses and explains the signs $>$ and $<$.

Signum majoritatis ut $a \mathrel{>\!\!\!-} b$ significet a majorem quam b.
  „   minoritatis ut $a \mathrel{-\!\!\!<} b$ „ a minorem „ b.

Sectio II. Æquationum Canonicarum ab Originalibus suis Derivatio sive Deductio. In this he shows the generation of equations from binomial factors or roots, and the deducing of Canonicals from their originals, e. g., if $a-b=0$, and $a-c=0$, then $aa-ab-ac=-bc$. In this and subsequent chapters he places the absolute term on the right-hand side of the equation.

Sectio III. Æquationum Canonicarum Secundariarum a Primariis Reductio per Gradus alicujus parodici Sublationem Radice Supposititia invariata Manente. This contains many examples of the removal of the second and third terms of an equation, up to equations of the fifth degree.

Sectio IV. Æquationum Canonicarum tam Primariarum quam Secundariarum, Radicum Designatio. In this Harriot shows how the roots can be determined from the form of an equation, but he considers positive roots only.

Sectio V. In qua Æquationum Communium per Canonicarum æquipollentiam, radicum numerus determinatur, namely, on the number of roots (positive roots) of common equations. This he determines by comparing the given equation with the like form among his Canonicals ; two equations having the same number of terms, and the same relations between the coefficients are called *Equipollents* *, and he declares that the number of positive roots in the proposed common equation will be the same as that used in compounding the canonical equation.

Sectio VI. Æquationum Communium Reductio per Gradus alicujus parodici Exclusionem et Radicis Supposititiæ Mutationem. In this chapter are many examples of reducing and transforming equations of the second, third, fourth degrees by multiplying the root of an equation in any proportion (Vieta) or increasing or diminishing the root by a given quantity (Cardan). In this section (p. 89) he calls

$$eee - 3 \cdot bbe = -ccc - 2 \cdot bbb$$

an *Impossible* equation because it has no affirmative root.

In the second part of the treatise, *Exegetice Numerosa,* Harriot gives numerical results for all kinds of equations up to the fifth degree.

These are the essential features of Harriot's treatise, from which it is clear that although it would be unjust to deny him a place in the history of science, it would be equally unfair to regard him as a pioneer in new branches of the subject.

Hutton (*Mathematical Dictionary,* p. 630) says that Doctor Zach maintains that from papers in his possession and printed in the Astronomical Ephemeris of the Royal Academy of Science, Berlin, 1788, it appears that Harriot was not less eminent as an astronomer and geometer than as an algebraist. He is a said to have applied the telescope to astronomical observations at the same time that Galilei did. Wood (Athen Oxon.) styled Harriot the " universal philosopher ", and Wallis, who had the highest regard for him, spoke of him in his *Algebra* (p. 126) as an " incomparable person ".

* " Duæ æquationes similiter graduatæ et similiter affectæ, quarum coefficiens ... et homogeneum datum unius coefficienti vel coefficientibus et homogeneo dato alterius in simplici inæqualitatis, majoritatis scilicet et minoritatis habitudine conformia sunt, *æquipollentes* in sequentibus appellandæ sunt ", Sect. v, p. 78 (p. 72 but follows p. 77).

APPENDIX I

In the Harleian Collection of MS. in the British Museum are three volumes of manuscripts from the pen of Harriot. *Tho. Harriot, Mathematical Papers* (Bibl. Harl. 6083). These seem to constitute a scientific scrap-book rather than a work written with a definite plan. Dissertations on square roots, annuities, " sayling in a greater circle ", an estimate of the number of persons who may inhabit the earth, and a diagram of the heart and the circulation of the blood are to be found side by side. On page 152 is a chapter on the Determination of the Roots of Cubic Equations. In the equation $aaa + 3BBa - CDE = 0$ he finds one root, and adds : " Reliquiæ duæ radices sunt impossibiles ". When he meets an equation which leads to imaginary roots he writes : " æquatio fictitia est ". There are also indications of an able treatise on Coordinate Geometry which includes the parabola, the ellipse, and the hyperbola. On page 242 is an attempt at a treatise on Solid Geometry, *Theoremata Phoranomica*. Then follows a *Traicte d'Algebre* (59 pages) in French. In the section dealing with cubic equations he writes :

$A3 + 0 + AGH = BCD$, 2 racines — and $1 +$

$A3 - 0 + AGH = BCD$, 3 racines$+$,

and he adds : " Lesquelles deux Equations se contrarient en deux Racines—et+monstrent, qu'il n'y a qu'une Racine +, et que les deux autres sont esvanouyes ".

Harriot employs the signs $>$, $<$, and ∞ for greater than, less than, and equal to, and he defines them thus :

" $>$ signifies more eligible,

$<$ signifies less eligible,

∞ signifies equallie eligible."

HEARNE, THOMAS (1678–1735).—This distinguished historian and antiquarian was educated at Oxford, where he received the degree of M.A. in 1703. He became Second Keeper of the Bodleian in 1712, and the next year he was offered the post of Librarian to the Royal Society, a post which he refused on account of his unwillingness to leave Oxford. He was deprived of his office at the Bodleian in 1716 because he refused to take the oath of allegiance to the Hanoverian Dynasty. On political grounds he refused the post of Custos Archivorum in 1726, preferring, to use his own words " a good conscience before all manner of preferment and worldly honour ".

He published *Reliquiæ Bodleianæ* (1703), *Collectanea* (1715), and many long English chronicles. His Diaries and Correspondence were printed by the Oxford Historical Society. Hearne maintained a constant correspondence with many literary men of his day. He had a most bellicose manner, often the result of a second-hand hearsay, and many distinguished scholars found themselves pilloried—and often unjustly— in his pages.

HÉRIGONE, PIERRE (dates not known).—A French mathematician of some eminence. In 1634 (2nd edition, 1644) he published a course of mathematics in Latin and French (*Cursus Mathematicus*) in six volumes, which bears evident marks of originality and ingenuity. The second of these contains a very able treatise on algebra, which is chiefly remarkable for the fact that in it, Hérigone attempted to reduce mathematical language to a universal symbolism equally intelligible to all nations. Besides $+$ for *plus* he uses $\sim$ for *minus*, $3|2$ for greater, $2|3$ for less, with several abbreviations of his own. In denoting powers he does not repeat the letters as did Harriot, but he subjoins the numerical exponents thus: $a2, a3 \ldots$ for a^2, a^3 etc. He also used numbers instead of letters to indicate roots, as $\sqrt{3}$ for cube root, instead of $\sqrt{C}$.

HEURAET, HEINRICH VAN.—Another Dutch geometer who developed the Cartesian Geometry. He contributed some papers on the nature of equations, but his chief glory was the fact that he made a successful attempt at curve rectification. Although his achievement was much later than Neile's, his method was quite independent of the Englishman's, and there is no reason to suppose that Heuraet had ever seen Neile's work.

HEVELIUS, JOHN (1611–87).—A famous Dantzig astronomer. His studies were very extensive, though mathematical subjects were his chief pursuit He built an observatory in order to carry out astronomical observations, the first results of which were published in 1647 under the title of *Selenographia, sive Lunæ Descriptio*. This treatise was remarkable for the accuracy of the observations of the moon in her different phases. Wallis was loud in his praises of this treatise *. Hevelius was the first to observe the phenomenon known as the Libration of the Moon. An account of this was published in a treatise *Epistola De Motu Lunæ Libratorio* (1654). His *Cometographia*, which appeared in 1668, occasioned the dispute with Hooke as to the respective advantages of plain and telescope sights. Hevelius wrote many other astronomical works, chief of which were the *Machina Cœlestis* (Pars Prior, 1673, Gedani) and the *Annus Climactericus* (1685). It is said that no one save Tycho Brahe had an observatory which could compare with his. This, added to a natural dexterity, made him one of the foremost astronomers of the seventeenth century. Hevelius' books and instruments were lost in a fire which broke out in 1679 and destroyed his observatory. Hevelius contributed many important papers to the *Transactions*. These are mainly on astronomical subjects, though one of them contains his observations on the variation of the magnetic needle at Dantzig (1670).

* Called it " Opus egregium .... Nec me sperantem fefellit expectatio "
M.S. Add. D. 105. April 3, 1649 (Bodleian).

APPENDIX I

HOBBES, THOMAS (1588–1679).—This well-known English philosopher was educated at Magdalen College, Oxford, where he graduated B.A. in 1608. After leaving Oxford, he travelled abroad, and became acquainted with Galilei, Gassendi, Mersenne, and Descartes, with all of whom he maintained a close correspondence on scientific subjects. On his return to England he became mathematical tutor to the Prince of Wales, afterwards Charles II, who, after the Restoration, granted him a life pension. He published *Leviathan* in 1651. He also wrote on a number of subjects, and as the opinions he expressed were for the most part heterodox he was frequently involved in acrimonious disputes, notably with Ward and with Wilkins. Not one of these disputes, however, was nearly as bitter as his protracted quarrel with Wallis.

In Metaphysics he was a thorough-going Nominalist. His other works include *De Cive* (1651), *De Corpore Politico* (1650), *De Homine* (1658), as well as the pseudo-mathematical writings mentioned in Chapter X. His works were published under the title : *The English Works of Thomas Hobbes* (Lond., 1839–45), by Sir William Molesworth.

Aubrey says of him : " His long life was that of a perfect honest man, a lover of his country, a good friend, charitable and obliging. He had a high esteeme for the Royal Societie . . . . and the Royal Societie (generally) had the like for him " (II, ii, 632).

HOLDER, WILLIAM (1616–98).—A celebrated Divine. He is described by Hutton (*Math. Dict.*) as a " general scholar, a very accomplished person, and a great virtuoso ". He was educated at Pembroke Hall, Cambridge, where he became M.A. and Fellow in 1640. He became Rector of Blechington in 1646, and was made D.D. in 1660. His promotion in the Church was rapid, and he became Canon of St. Paul's, and Sub-Dean of the Chapel Royal in 1674.

Holder was elected to a Fellowship of the Royal Society in 1663. It was about this time that he greatly distinguished himself by teaching to speak a youth born deaf and dumb. He subsequently outlined his methods in a treatise *Elements of Speech* (1669), which was described by Aubrey as a " most ingeniose and curious discourse, and untouched by any other ; he was beholding to no author ; did only consult with nature ". The same writer went on to say : " Dr. Jo. Wallis unjustly arrogated to himself the glory of teaching y sayd young gentleman to speake, in the *Philosophical Transactions*, and in Dr. Rob. Plott's *History of Oxfordshire* " *.

He is said to have assisted in the education of Christopher Wren, whose sister he married. In addition to the above-mentioned treatise, Holder also wrote : *A Discourse concerning Time, with Applications to the Natural Day, Lunar Month, and Solar Year* (1694) ; *Principles of Harmony* (1694).

Aubrey summed up his character in the words : " If one would goe about to describe a perfect good man, would draw this Doctor's character ".

* Aubrey II, ii, 398/9.

HOOKE, ROBERT (1635–1702/3).—Is described by Aubrey (II, ii, 406) as " a person of great vertue and goodness .... He is certainly the greatest mechanick this day in the world ". After being educated at Westminster he went to Oxford, where he was instructed in astronomy by Seth Ward. He became keenly interested in experimentation, and frequently he assisted Dr. Thomas Willis in his chemical experiments and Boyle with his air pumps. In 1663 he was elected to a Fellowship to the Society, and on January 11, 1664/5 he was appointed Curator to the Society for life with a salary of £30 per year. He was without doubt one of the most original and versatile scholars of the day. He was endowed with an inexhaustible fertility for devising experimental proofs, and he employed his talents unwearyingly in the service of the Society. His experiments covered a wide range, and the Record Books of the Society testify to the eagerness with which he hurried from one enquiry to another. In 1665 he became Gresham Professor of Geometry, and in the same year he published his *Micrographia*. He expounded the true theory of Elasticity, and in some measure he anticipated the Inverse Square Law of Newton. He also has some important astronomical discoveries to his credit; he made the earliest attempt at a telescopic determination of the parallax of a fixed star; he constructed the first Gregorian Telescope, as well as other instruments for making observations both on land and on sea. It is said (*e. g.*, Hutton, *Math. Dict.*) that Hooke was of a jealous and ill-tempered nature, " so envious and ambitious that he would fain have been thought the only man who could invent and discover anything ". From his Diary, recently published, it seems not unreasonable to assume that his vituperation was often the result of prolonged ill-health. He managed his quarrel with Hevelius with such bitterness as to be universally condemned. In 1671 he attacked Newton's *Theory of Colours*, and soon afterwards he quarrelled with Oldenburg.

Hooke does not appear to have appreciated the greatness of Wallis, and the following quotation from the *Diary* is typical of many: " Oct. 26 ", he wrote in 1678, " Left 3 sheets with Dr. Wallis. He blundered .... To Sir J. Hoskins' chamber; Wallis read papers but understood them not, fain'd sleep, denyed the necessity of 4plex power for 2ble velocity ".

On the death of Oldenburg (1677), Hooke was appointed to his post. He did not, however, continue the publication of the *Transactions*. His chief writings include : *Lectiones Cutlerianæ* (1674), *Micrographia* (1665), *Philosophical Collections*. His posthumous works were published by Richard Waller, Secretary to the Royal Society, in 1705, and by Derham in 1726. Hooke was also Surveyor to the City of London, and he designed a number of important buildings such as Montagu House, Bethlehem Hospital, etc.

Birch opens his *Life of Hooke* with the eulogy : " Dr. Robert Hooke is generally allowed to have been one of the greatest promoters of Experimental Natural Knowledge, as well as Ornaments of the seventeenth century, so fruitful of great genius ".

HORROX (HORROCKS), JEREMIAH (1619–41).—Was a sizar at Emmanuel College, Cambridge from 1632 to 1635. At an early age he began to apply himself diligently to astronomical research, and he is said to have observed a partial eclipse of 1639 with a half-crown telescope. He predicted and observed a Transit of Venus (Nov. 24, 1639) before the age of twenty, and soon afterwards he detected the long inequality of Jupiter and Saturn. His most famous work, *Venus in Sole Visa*, was first printed by Hevelius at Dantzig in 1662. Another important treatise, *New Theory of Lunar Motions*, was highly appreciated by Newton, who spoke of its author as a genius of first rank. Horrox turned his attention to the phenomena of the Tides in 1640, and he is said to have identified solar attraction with terrestial gravitation.

His works were published posthumously by the Royal Society in 1672 and 1678, Wallis being largely instrumental in seeing them through the Press.

HUDDE, JOHN (1633–1704).—Was distinguished alike in politics and in mathematics. He applied himself particularly to the analysis of equations, and we have from him two fragments, both inserted in Schooten's Commentary (*Jo. Huddenii : de Reductione Æquationum et de Maximis et de Minimis. Epistle* 11, Amsterdam, 1658). These contain a very ingenious method of discovering whether an equation has equal roots, and for finding these roots. In the same work Hudde perfected the method of Descartes and Fermat for drawing tangents to curves, and for discovering maxima and minima. Hudde planned a great work *De Natura, Reductione, Determinatione, Resolutione, atque Inventione Æquationum*. Portions of this also appeared in the Introduction to Van Schooten.

HUYGENS, CHRISTIAN (1625–95).—First gave indication of his mathematical genius when he published his *Theoremata de Quadratura Hyperbolæ, Ellipsis, et Circuli* (1651), in which he showed the fallacy in a system of quadratures proposed by Grégoire de Saint Vincent. This Essay was followed by tracts upon the Quadrature of the Conics and the approximate rectification of the Circle (*De Circuli Magnitudine Inventa*, 1654). In 1654 he directed his energies to improvements in the telescope, and by means of his improved instrument he was able to solve a number of astronomical questions, *e. g.*, the nature of Saturn's Rings. These were described in his *Systema Saturninum* (1659). He also devoted considerable attention to devising exact methods of measuring time, and his researches led to the invention of the pendulum clock, which he described in his *Horologium Oscillatorium sive de Motu Pendulorum* (Paris, 1673). This treatise also contains a complete account of the descent of heavy bodies either vertically or down smooth curves. The theory of evolutes is developed, and in discussing the cycloid, he showed that this curve was tautochronous. He investigated the Compound Pendulum, and he showed that the Centres of Suspension and of Oscillation were interchangeable.

In 1663 he was elected a Fellow of the Royal Society. Five years later, at the invitation of the Society, he attacked the problem of Impact of Bodies, and proved that the momentum in a certain direction before collision was equal to the momentum in that direction after collision. About this time he became involved in a dispute with James Gregory concerning the quadrature of the hyperbola. This grew to be very acrimonious, especially on the side of Gregory, whose defence was inserted at his own request in the *Philosophical Transactions*. Huygens was admitted a member of the French Académie des Sciences in 1666, and eventually he became one of its most distinguished members.

Apart from his mathematical researches, Huygens' most outstanding work was his exposition of the Undulatory Theory of Light. This appeared in 1690. On this theory he deduced by the application of geometry the laws of reflection and refraction of Light, and he also explained the phenomenon of double refraction. Like Wallis, he also did valuable experimental work.

Sprat eulogises him in the words : " This gentleman has bestowed his pains on many parts of the speculative and practical mathematics with wonderful success, and particularly his applying the motion of pendulums to clocks and watches was an excellent invention. For thereby there may be found a means of bringing out the measures of time to an exact regulation " *.

KEPLER, JOHANN (1571–1630).—A celebrated German astronomer. He was educated at the University of Tübingen, where he assimilated the Copernican doctrines. His work, *Mysterium Cosmographium*, brought him to the notice of Tycho Brahe and Galilei, and he became assistant to the former at Prague. On the death of Tycho (1601) he became Imperial Mathematician to Rudolph II, and he inherited most of Tycho's observations. In 1604 he published *Astronomiæ Pars Optica*, in which he established his laws of elliptical orbits and of equal areas. This was followed in 1609 by *Astronomia Nova* (Prague), in which he showed the connection between the tides and lunar attraction.

In his *Nova Stereometria Doliorum* (Linz, 1615) he evolved rules for determining the areas and volumes of figures bounded by curved surfaces. This work had a profound influence upon Cavalieri and on Wallis.

Other works of his are :

(1) *Ad Vitellionem Paralipomena*. 1604.

(2) *Dioptrice*. Ausburg, 1611.

(3) *De Harmonica Mundi*. Ausburg, 1619. This contains the third law which states the connexion between planetary distances and periods.

* Sprat, pp. 177, 178.

APPENDIX I

(4) *Epitome Astronomicæ Copernicæ.* (1618–21.) This contains a lucid exposition of the Copernical Hypothesis.

(5) *Tabulæ Rudolphinæ.* Ulm, 1627.

KERSEY, JOHN (1616–90).—Was encouraged in the study of mathematics whilst quite young. He wrote *The Elements of that Mathematical Art, commonly called Algebra* (Two Vols.), and was persuaded by Collins to publish. It appeared in the years 1673 to 1674, and is remarkable for the clearness of its exposition. This gave its author a great reputation. Both Wallis and Collins spoke in terms of the highest praise of this work, and it received great commendation in the pages of the *Transactions*. He edited the *Arithmetic* of Edmund Wingate, 1650–53.

LALOUÈRE (LOUBÈRE), SIMON (1600–64).—Published a work *Elementa Tetragonismica seu Demonstratis Quadraturæ Circuli et Hyperbolæ ex Datis ipsorum Centris Gravitatis* in 1651 (Tolosæ). He also wrote a work on the Cycloid (*Geometrica Promota de Cycloide,* 1660), which was described by Montucla as "un géométrie marchant toujours par les routes embarrassées". He is of interest on account of the fact that, like Wallis, he attempted to solve Pascal's Prize Questions on the Cycloid. His solutions were much inferior to those submitted by Wallis.

LANGBAINE, GERARD (1609–58).—This celebrated Divine was Wallis's predecessor as Custos Archivorum to the University of Oxford. He entered Queen's College, Oxford, in 1625, and became M.A. and Fellow in 1633, Provost of Queen's College, and D.D. in 1646. He was a zealous Royalist and supporter of Episcopacy. He wrote many literary and political pamphlets and he has left many collections of notes in the Bodleian. According to Wood, Langbaine "made several Catalogues of Manuscripts, nay, and of printed books too, in order as we suppose for an Universal Catalogue in all kinds of learning"*. The same writer describes him as a "great ornament of his time to this University".

LEOTAUD, VINCENT (1595–1672).—A French geometer, who taught mathematics at Lyons. He was the author of several treatises which in their day attracted the notice of mathematicians. He wrote *Geometriæ Practicæ Elementa, ubi de Sectionibus Conicis habet quædam Insignia* (Dolæ, 1631), and *Examen Circuli Quadraturæ quam . . . . Gregorius a Sancto Vincentio exposuit* (Lugduni, 1654), which was a refutation of a work published a few years earlier by Grégoire de Saint Vincent. He also wrote *Institutionum Arithmeticarum Libri* iv, (Lyons, 1660).

* Wood, *Ath. Oxon.*, II, col. 141.

MERCATOR (KAUFMANN), NICHOLAS (1640–87).—A native of Holstein. He spent most of his time in England, and he became one of the first members of the Royal Society. He had a conspicuous genius for mathematical sciences, and in 1668 he published his chief work, *Logarithmotechnia* (Lond., 1668 and 1674). This contained his famous series :

$$\log(1+x) = x - \tfrac{1}{2}x^2 + \tfrac{1}{3}x^3 - \tfrac{1}{4}x^4 +,$$

which he proved by writing the equation of the hyperbola in the form

$$y = \frac{1}{1+x} = 1 - x + x^2 - x^3 +,$$

to which Wallis's methods of quadrature could be applied. Mercator has frequently been accused of borrowing the inventions of others without making due acknowledgments.

In addition to the above, he wrote *Cosmographia* (Dantzig, 1651); *Astronomia Spherica* (1651); *Hypothesis Astronomica Nova* (Lond., 1664). He seems to have dabbled in astrology, which he tried to reduce to rational principles, and Montucla wrote concerning him : " On est fâché d'apprendre qu'après la mort de Mercator, on trouva parmi ses papiers un Traité d'Astrologie, de sa main " *.

MERRETT, CHRISTOPHER (1614–95).—A physician of repute. He became M.D. at Gloucester Hall in 1643, F.R.C.P. in 1651, and Gulstonian Lecturer in the following year. He published many works on Natural History and Medicine, and he contributed many papers to the *Transactions*. He helped in the formation of the Royal Society, and became a Fellow upon its Incorporation.

MERSENNE, MARIN (1588–1648).—A Franciscan Friar who studied at La Flèche, where he met and became intimate with Descartes, his junior by a few years. He rapidly acquired a commanding position in the world of science, and he maintained a close correspondence with many learned men of his time, being in France what Collins was in England. Like Collins, he urged writers to publish their works, and thus, through his encouragement, science has been enriched with many works which might otherwise have been lost. Many French writers, *e. g.*, Fermat, claim that the cycloid was invented by him, but this opinion was stoutly challenged by Wallis, who declared that this curve was known to Cardinal Cusa about 1450 †. Mersenne has some interesting problems on the Theory of Numbers, but his best work was in *Physics*. He is chiefly remembered by :—

(1) *Questiones celeberrimæ inouies ou Récréations des Scavans*, 1634. These deal mainly with the philosophy of mathematics.

(2) *Harmonicorum Libri xii* (Paris, 1648).

* Montucla, ii, p. 356.
† *Phil. Trans.*, vol. iv, p. 169.

APPENDIX I 205

(3) *Cogitata Physico-Mathematica*, 1634.

(4) *L'Harmonie Universelle* (Paris, 1626/7).

(5) *La Vérité des Sciences* (1625).

In addition, he translated many of Galilei's works into French (1634) and thus helped to popularize them on the Continent. He was one of the most distinguished members of the Académie des Sciences, mainly on account of his intelligent appreciation of the work of others and his skill as an experimenter. He is said to have resigned all civil preferment in order to devote himself wholly to the development of learning.

Many of his letters are to be found amongst those of Descartes. His correspondence was published in 1932 by Paul Tannery (Paris).

MORAY (MURRAY), SIR ROBERT (?-1673).—Was one of the founders of the Royal Society. He was knighted by Charles I, and he successfully negotiated on his behalf between France and Scotland, and unsuccessfully planned his escape. He was learned in Geology, Chemistry, and in Natural History, and he contributed several papers to the *Transactions*.

He was a Charter Member of the Royal Society, and at several of its meetings before its Incorporation he occupied the Chair. According to the Record Book, the President was elected monthly; Moray was elected on March 6th, 1661, and re-elected April 10th, 1661. He was again President on August 28th, 1661, and on February 5th, June 11 and 18th, and July 2nd, 9th, 16th, 1662. Wilkins presided on May 21st, 28th, June 4th, 1662, and Boyle on June 25th.

MORLAND, SIR SAMUEL (1625–95).—A diplomatist and a mathematician He was educated at Winchester, and at Magdalen College, Cambridge where he became a Fellow in 1649. According to Hutton (*Math. Dict.*), he was also an ingenious mechanist and philosopher. He joined Charles II at Breda, to whom he afterwards became Master of Mechanic, and he invented several useful machines, *e. g.*, a fire-engine, a speaking trumpet, and a capstan for heaving up anchors. He published a tolerably good treatise on Arithmetic in 1674, and he has three papers in the *Transactions*.

MYDORGE, CLAUDE (1585–1647).—Is notable, not only as a geometer, but also as a friend of Descartes. In 1631 he published an Introduction to *Dioptrics and Catoptrics*, and Two Books on Conic Sections (*Prodromi Catoptricorum et Dioptricorum*, etc., Paris, 1631). His *De Sectionibus Conicis* (Paris, 1644) earned for him high praise. Chasles says: "Mydorge .... eut le mérite d'être le premier en France qui écrivit un *Traité des Sections Coniques*, et qui entreprit de simplifier les démonstrations des anciens, et d'aller au delà de ce qu'ils avaient fait sur ce sujet"*. His work was much more concise than that of the ancient

* Chasles *Aperçu Historique*, p. 88.

geometers, in fact, it was claimed that in this treatise Mydorge could demonstrate in one proposition properties for which the ancient geometers would have required three. He also wrote *Examen du Livre des Récréations Mathématiques* (1639 and 1648).

NAPIER (NEPER), JOHN (1550–1617).—Baron of Merchiston, was educated at St. Andrews. He is best remembered by his *Mirifici Logarithmorum Canonis Descriptio* (Edin., 1614), in which he propounded his discovery, as well as the use of, logarithms. Three years later he published *Rabdologiæ, seu Numeratio per Virgulas* (Edin., 1617), in which he explained various ingenious methods of calculating by means of Napier's "rods" or "bones".

Napier is generally considered to be the inventor of the modern system of writing decimal fractions, and in the *Rabdologiæ* he freely uses the comma and the full-stop to indicate the position of the decimal point.

The following works were published after his death :

(1) *Mirifici Logarithmorum Canonis Constructio* (1619).

(2) *Arithmetica Logarithmica*. Lond., 1624.

(3) *De Arte Logistica*. Edin., 1842.

NEILE, WILLIAM (1637–70).—This distinguished mathematician was a gentleman commoner at Wadham College, Oxford. Here he came into contact with Ward and with Wilkins, both of whom encouraged his genius for mathematical studies. Neile's chief title to fame lies in the fact that by his rectification of the semi-cubical parabola in 1657 he first found the length of a curved line. Neile became a Fellow of the Royal Society in 1663. In 1669 he presented to the Society his *Theory of Motion*, which greatly enhanced his reputation. He also wrote a slight work, *De Motu*, Lib. 1. Further expectations of his genius were cut short by his early death.

OLDENBURG, HENRY (1615–77).—Was born and educated at Bremen. He came to England in 1640, as Consul for his country. He went to Oxford in 1656 and there he made the acquaintance of Boyle and other members of the group which presently gave rise to the Royal Society. With Wilkins he was made first Secretary to the Society (1663), and he applied himself with marked diligence to the duties of that office. In 1664 he began the publication of the *Philosophical Transactions*. In order to discharge this task more efficiently he maintained a correspondence with more than seventy men of learning on a variety of subjects from all over the world. He was a constant correspondent of Boyle, and he translated many of his works into Latin. About 1665 he was drawn into a dispute with Hooke, who complained that the Secretary had not done him justice with respect to the invention of spiral springs for pocket watches. This dispute was carried on with considerable warmth on both sides, but it was at last terminated to the honour of Oldenburg, the Council approving his action. In the 'Hooke Diary', the author stigmatises Oldenburg as "a lying dog, treacherous, and a villain".

Oldenburg continued to publish the *Transactions* until No. 136 (January, 1677). In addition, he published some twenty tracts, chiefly on theological and political subjects, and one of the latter led to his imprisonment in 1667.

OUGHTRED, WILLIAM (1574/5–1660).—An eminent English mathematician and Divine. He was educated at King's College, Cambridge, where he became a Fellow. In 1603 he was made Rector of Aldbury, near Guildford, where he led a long and studious life. In 1628 he became tutor to Lord William Howard, son of the Earl of Arundel, and his *Clavis* was drawn up for the use of his pupil. Oughtred kept up a constant correspondence with many eminent scholars, mainly on mathematical subjects. His chief contributions to Science were :—

(1) *Arithmeticæ in Numeris et Speciebus Institutio, quae . . . . totius Mathematicæ quasi Clavis est* (1631). This treatise he intended should serve as a key to Mathematics. It was afterwards reprinted with a tract on the Resolution of Equations in Numbers, in 1648, with the title *Clavis Mathematicæ*.

(2) *Circles of Proportion* (1633 and 1660).

(3) *Trigonometria* ; A treatise on Trigonometry (1657). In this he employed the trigonometrical abbreviations now in use.

(4) A large number of papers on mathematical subjects. These were printed at Oxford in 1677, under the title *Opuscula Mathematica hactenus Inedita*. These include some writings on mechanics.

Oughtred's style and manner were very concise and obscure, and his rules so involved in a maze of symbols and abbreviations as to render his mathematical writings very difficult to read. He employed no less than 150 symbols. These include :—

$\times$ for multiplication.

: : for proportion.

$\stackrel{..}{\cdot\cdot}$ for continued proportion.

⊐ for " greater than ".

⊏ for " less than ".

Oughtred's work earned the warmest commendation from many distinguished men of science. Wallis dedicated his *Arithmetica Infinitorum* to him. Boyle wrote : " The *Clavis* doth much content me ". Newton, Twysden and others assign to him a high place in mathematics. " The best Algebra yet extant is Oughtred's "*.

* *Life of Locke*, ed. King, i, 227.

Aubrey says, however, that he was more famous abroad for his learning than at home, and that several great men came over to England for the purpose of meeting him *. This is no doubt an exaggeration; on the other hand, it is probably quite true as has been suggested, that Seth Ward learned his mathematics from him.

Oughtred was a staunch Royalist, and he is said to have died in a transport of joy on hearing the news of the Restoration of Charles II.

PELETIER, JACQUES (1517–82).—Born at Mans and resided at Paris, Bordeaux, Lyons, and Rome. He wrote a treatise on Algebra with the title *De Occultate Parte Numerorum, quam Algebram vocant*, Libri duo (Paris, 1554), which in its symbolism shows both German and Italian influences. He indicates powers and roots by numbers as did Stifel, but he follows the Italian mathematicians by his use of p and m for plus and minus. He died in Paris.

PELL, JOHN (1610/11–85).—This well-known mathematician was educated at Trinity College, Cambridge, where he became M.A. in 1630. He went abroad on leaving college, and became Professor of Mathematics at Amsterdam in 1643, whence he removed three years afterwards to Breda. He returned to England in 1652, and was employed by Cromwell as a diplomatist in Switzerland from 1654 to 1658. Here he became intimate with Rahn. After the Restoration he took Orders, and in 1663 he was made D.D. His mathematical reputation was great, but he accomplished little, and he has left nothing of moment. The division sign $\div$, however, is due to him and, according to Wallis †, so is the "Note of Illation" $\therefore$ *Ergo*. The indeterminate equation $ax^2 - 1 = y^2$ bears his name, although his connection with it consists of the publication of the solution of it in his edition of *Branker's Translation of Rhonius' Algebra*. His chief mathematical works were :—

(1) *An Idea of Mathematicks* (Lond., 1650).

(2) *A Table of* 10,000 *Square Numbers*, etc. (1672).

(3) *An Inaugural Oration at his entering upon the Professorship at Breda*.

(4) *An Introduction to Algebra; translated out of the High Dutch into English, by Thomas Branker, much altered and augmented by Dr. Pell* (Lond., 1668).

Pell was a great admirer of Harriot, and he remarked that had Harriot published all that he knew in algebra he would have left little of the chief mysteries of that art unhandled.

In his *Algebra* (p. 198) Wallis relates a story quoted by Pell in which he describes how Roberval, on being shown a copy of Harriot's treatise, exclaimed : " Il l'a vue, il l'a vue ! " by which he meant that Descartes had not only seen, but had also borrowed many of the improvements of Algebra which are to be found in his pages.

* Aubrey, ii, p. 471.
† *Algebra*, p. 219.

APPENDIX I

PETTY, SIR WILLIAM (1623–87).—Is described by Hutton as " a singular instance of a universal genius ". His chief fame is that of a political economist. He studied on the Continent at Caen, and he became intimate with Hobbes. In politics he espoused the side of the Parliamentarians. In 1651 he was made Oxford Professor of Anatomy.

He executed for the Commonwealth the " Down Survey " in Ireland, the first attempt on a large scale at carrying out a social survey systematically. This survey was based upon a collection of social data which entitles him to be considered a pioneer in the science of comparative statistics. On the return of Charles II he acquiesced in the Restoration and was knighted. In 1662 he became one of the original members of the Royal Society and a member of its Council. He published many treatises on Economics, e. g., *Treatise on Taxes and Contributions* (1662), and *Essays concerning Multiplication of Mankind* (1686), as well as several contributions to the *Transactions*. He was President of the Dublin Philosophical Society in 1684, and according to Aubrey, he drew up a catalogue of " mean, vulgar, cheap and simple experiments for the infant Society ".

RAHN, JOHANN HEINRICH (1622–76).—Was a native of Zurich, and the son of the Burgomaster there. He wrote *Algebra Speciosa*, extant only in manuscript, which dealt mainly with the solution of difficult equations. He is, however, best remembered by his *Teutsche Algebra* (Zurich, 1659), which was remarkable for the amount of symbolism employed in its pages. This was translated by Branker and published in London, 1668, with the title : *An Introduction to Algebra, by Rahn. Translated by Th. Branker; much Augmented by Dr. Pell.* Lond., 1688.

ROBERVAL, GILES PERSONNE (1602–75).—Was a mathematician of repute. Born at Beauvais, he came to Paris in 1627, and there he met several distinguished scholars, including Mersenne. He became Professor of Mathematics there, and on the institution of the Académie des Sciences in 1665 he became one of its first members. Roberval investigated the nature of tangents to curves, solved problems on the cycloid, and made important observations in mechanics. In his hands the Method of Indivisibles, employed by Cavalieri, was developed and greatly improved. He also made observations on the pressure of the air, and these are included in his *Essais Mécaniques*. He appears to have been something of a pedant, and his violent letter to Torricelli lends support to that belief. He quarrelled with Descartes as well as with others, and according to his fellow-countrymen Montucla, " il montra dans la plûpart de ces démêlés beaucoup plus de passion que de savoir et d'amour pour la vérité ". That same writer does not regard him as a very original and penetrating thinker, for he adds : " M. de Roberval n'eut jamais l'art d'exposer ses idées avec netteté et précision " *.

* Montucla, ii, 50/51

In addition to the *Essais*, Roberval also contributed the following tracts, which are to be found in volume vi of the Memoirs of the Académie des Sciences :—

(1) *Sur la Composition des Mouvements.* Anc. Mém., Paris, T. vi.

(2) *De Recognitione Æquationum.*

(3) *De Geometrica Planarum et Cubicarum Æquationum Resolutione.*

(4) *Traité des Indivisibles.*

ROOKE, LAWRENCE (1622–62).—Was an English astronomer and geometer. He was educated at Eton and King's College, Cambridge, where he obtained the degree of M.A. in 1647. In 1650 he went to Oxford in order to be near Ward and Wilkins, and to attend Boyle's chemical experiments. On the death of Foster (1652) he was chosen Astronomy Professor at Gresham College, but five years later he was permitted to change his Astronomy Professorship for that of Geometry.

Rooke was very zealous in furthering the movement which ultimately led to the formation of the Royal Society, but he died just before its Incorporation. He seems to have enjoyed the greatest respect amongst his contemporaries ; Wallis dedicated his *Conic Sections* to him ; Barrow dwelt upon his high character in his Oration at Gresham College. Ward and Wilkins spoke of him as a man of " profound judgment, vast comprehension and solid experience ", and even Hooke wrote : " I was never acquainted with any person who knew more and spoke less, being indeed eminent for the knowledge and improvement of Astronomy ". His chief writings were :

(1) *Observations on the Comet of December 1662.*

(2) *Directions to Seamen bound for Long Voyages* (*Phil. Trans.*, Jan. 1662).

(3) *A method of Observing the Eclipses of the Moon* (*Phil. Trans.*, Feb. 1666).

SAINT-VINCENT, GRÉGOIRE DE (1584–1667).—A Flemish geometrician, who studied mathematics under Clavius. As a writer he is very diffuse, and he is chiefly remembered by his attempts at circle squaring. Montucla observes that " no one ever squared the circle with so much ability or (except for his principle object) with so much success ". The fallacy in these attempts was pointed out by Huygens.

His chief mathematical treatise was his *Opus Geometricum Quadraturæ Circuli et Sectionum Coni* (Antwerp, 1647). This contained a great number of new theorems on the properties of the circle and the conic sections, geometrical progressions, and volumes of solids of revolutions. One of these led to the expansion of $\log(1-x)$, in ascending powers of x. In an earlier work (*Theoremata Mathematica*, 1624) Saint-Vincent gave a concise account of the Method of Exhaustions which is applied to several quadratures, notably that of the hyperbola. Chasles entertained a high regard for the work of Saint-Vincent, as also did Montucla. The latter speaks of the *Opus Geometricum* as " un vrai trésor, une mine riche de vérités géométriques et de découvertes importantes et curieuses "*, and Chasles asserts that " si les travaux de Grégoire de Saint-Vincent n'ont point été cultivés comme ils étaient dignes de l'être, la cause en est due, sans doute, à l'invention presque contemporaine de la Géométrie de Descartes, et de l'Analyse Infinitésimale, qui ont tourné toutes les méditations vers le calcul " †.

SCHOOTEN, FRANCIS, VAN (1584–1646).—A native of the Netherlands, was Professor of Mathematics at Leyden. He was a great admirer of Descartes, and he brought out in 1649, and again in 1659, an edition of his *Géométrie*, together with the notes thereon by De Beaune. In 1646 he published a treatise, *De Organica*, in which he expounded the different methods of describing the Conic Sections. Two years later he brought out his *Principia (Principia Matheseos Univers.* 1651). But his chief work was his *Exercitationes Mathematicæ* (Leyden, 1657), in which he applied the new analytical geometry to the solution of many interesting and difficult problems.

SLOANE, Sir HANS (1660–1753).—A distinguished physician and naturalist who studied at Paris. He was elected a Fellow of the Royal Society in 1685, and he became its Secretary in 1693. He held this post until 1712, and during his tenure of office he revived the publication of the *Transactions*. He became President of the Society in 1727, and he continued to hold that office until 1741. He was also made a foreign member of the Academies of Science at Paris, at St. Petersburg, and at Madrid. He was also President of the Royal College of Physicians from 1719 to 1753. His collections were purchased by the nation and placed in Montagu House (afterwards the British Museum) in 1754.

DE SLUZE, RENÉ FRANÇOIS WALTHER (SLUSIUS) (1622–85).—Described by Wallis in his *Algebra* as " a very accurate and ingenious person ". He was a Canon of Liége, where he acquired great celebrity by his knowledge of mathematics and physics. The Royal Society elected him one of its members, and inserted several of his works in the *Transactions*, e. g., his method of drawing tangents to geometrical curves (*Phil. Trans.*, vol. vii, 1672). He also published a work *Mesolabium et Problemata Solida*, 1668, and later, a new edition of the same, with the addition of a miscellaneous collection of important papers relating to Spirals, Centres of Gravities, Maxima and Minima, all of which showed great skill and industry.

SMITH, ROBERT (1624–1716).—A learned Divine. He was a student of Christ Church, Oxford, from 1651 to 1655, and he became M.A. in 1657. He was Public Orator at Oxford from 1660 to 1677. In 1673 he was offered the See of Rochester, but he declined.

* Montucla, ii, p. 79.
† Chasles, *Aperçu Historique*, p. 91 (footnote).

STEVINUS, SIMON (1548–1620).—Though chiefly remembered by his contributions to mechanics and hydrostatics, Stevinus made notable contributions to algebra, and he was responsible for some distinct progress in notation. He used exponents to indicate powers; thus he wrote $3x^2-5x+1$ as 3 ②$-$5 ①$+$1 ⓪. He had also a similar notation for decimals of which he was quite possibly an independent inventor. He wrote 346·735 thus 346 ⓪ 7 ① 3 ② 5 ③.

His reputation rests on his *Statics and Dynamics*, which was published in Flemish in 1586. In this he enunciated the Principle of the Triangle of Forces. In Hydrostatics he discussed the so-called hydrostatic paradox. His *Œuvres Mathématiques* was published at Leyden in 1634.

STUBBE, HENRY (1632–76).—Physician and author, was educated at Christ Church, Oxford (M.A. 1656), where he was highly esteemed for his learning. He was, however, expelled from his college in 1657 for writing against the Clergy and the Universities. " For a pestilent book of this kind ", says Wood, " the Dean of Christ Church ejected him from his studentship " *. Later he was imprisoned for writing a pamphlet denouncing the marriage of James, Duke of York, to Mary of Modena. He served with the Parliamentary Forces from 1653 to 1655.

Stubbe is described by Wood as " the most noted Latinist and Grecian of his age . . . . a singular Mathematician, and thoroughly read in all political matters " †. He was intimately acquainted with Hobbes. He wrote *The Commonwealth of Oceana, put in a Ballance, and found too Light* ", 1660 (An Account of the Republic of Sparta), and *A Severe Enquiry into the late Oneirocritica, published by John Wallis, or An exact Account of the Grammatical Part of the Controversy betwixt Mr. Thos. Hobbes and J. Wallis, D.D.* (Lond. 1657).

TANNER, THOMAS (1674–1735).—A noted antiquary, who afterwards became Bishop of St. Asaph. He was educated at Queen's College, Oxford, where he was made M.A. in 1696. In the same year he became Fellow of All Souls College, Oxford. In 1710 he was made D.D. and was raised to the Bishopric in 1732. He wrote *Notitia Monastica* (Oxford, 1695), or *A Short History of Religious Houses in England and Wales*; *Bibliotheca Britannico-Hibernica* (1748), an account of all the authors who flourished within the three kingdoms at the beginning of the seventeenth century. He was one of the literary executors of Wood, and he published, with some modifications, the continuation of his *Athenæ Oxonienses*. He amassed a great quantity of material for a History of Wiltshire, which is at present in the Bodleian.

* Wood, *Fasti Oxon*, ii, 175.
† Wood, *Athen. Oxon.*, 414.

APPENDIX I

TORRICELLI, EVANGELISTA (1608–47).—Was a pupil of Galilei. At the age of twenty, he was sent to Rome to study mathematics under Castelli, Professor of Mathematics there. After the death of Galilei he went to Florence, where he applied himself with intense zeal to the study of mathematics, physics and astronomy. His best known work is his *Opera Geometrica* (Florence, 1644), in which he extended the principles of dynamics to liquids. He also wrote on the quadrature of the Cycloid and the Conics, which occasioned disputes with Roberval. He greatly improved the art of making microscopes and telescopes, but it was by his work on the barometer that he was best remembered. His work was eagerly studied by the Royal Society.

TURNER, PETER (1586–1652).—According to Wood, Turner was one of the first scholars to serve King Charles I. He was taken prisoner by the Parliamentary Forces at Edge Hill and removed to Northampton, where he suffered great privations and died soon after. He was styled by Archbishop Usher as " Savilianus in Academia Oxoniensi Matheseos Professor eruditissimus ".

He was Gresham Professor of Geometry from 1620–31, and Savilian Professor 1631–48. He assisted in revising the University Statutes.

Ward says that " he was much beloved of Archbishop Laud, and so highly valued by him that he would have procured him to be one of the Secretaries of State, or Clerks of the Privy Council, but being wedded to his College and a studious Life, he denied these and other honourable and beneficial places " *.

UBALDO, GUIDO DEL MONTE (1545–1607).—A nobleman of Pisa, who wrote on mechanics. He studied at Padua, and afterwards fought against the Turks. He tried to reduce all his problems to the principle of the Lever, and according to Cantor he conceived the principle of Virtual Velocities. " In seiner Mechanik von 1577 ist das Gesetz enthalten, dass Last und Kraft zu einander im umgekehrten Verhältnisse der Räume stehen, welche sie in derselben Zeit durchlaufen ", *i. e.*, his mechanics enunciated the law that weight and force are inversely proportional to the distances traversed in the same time. His chief work is his *Mechanicorum* (Pisa, 1577). Ubaldo also wrote *Theorica Planisphæriorum* (1579), which contains some interesting mathematical constructions.

VARRO, MICHAEL (died 1586).—He wrote *Tractatus de Motu* (Geneva, 1584), in which he enunciated the principle that the velocity acquired by a falling body must be proportional to the space described since the motion began. Quite possibly he had the notion of the Composition of Forces (Whewell, p. 8).

* Ward, p. 134.

VIETA, FRANCISCUS (FRANÇOIS VIÈTE) (1540–1603).—A distinguished French mathematician. His works are replete with marks of originality and erudition. Chief among these is the *In Artem Analyticam Isagoge* (1591), which is usually considered the earliest work on symbolic algebra. In this he denoted known quantities by consonants, and unknown quantities by vowels, and he wrote the powers of his quantities by adding the words quadratus, cubus, etc. He also indicated how from a given equation another could be formed whose roots were equal to those of the original increased by a given quantity, and he employed this method to eliminate the coefficient of x in a quadratic, and of x^2 in a cubic. A later work, *De Æquationum Recognitione et Emendatione* (1615) consists of fourteen chapters leading to the resolution of equations by a rule similar to Cardan's. Chapter XIV contains four theorems in which the relations between the roots and the coefficients are clearly set out. From this it is clear that Vieta was acquainted with the fact that the first member of an algebraic equation $f(x)=0$ could be resolved into linear factors. But he dealt with positive roots only.

All his mathematical works were published by Van Schooten in 1646, with the title *Opera Mathematica*. In addition to the above-mentioned works, these include :

(1) *Ad Logisticem Speciosam Notæ Priores.*

(2) *De Numerosa Potestatum ad Exegim Resolutione* (Paris, 1650).

(3) *Supplementum Geometriæ* (1593).

VIVIANI, VINCENZO (1621–1703).—A celebrated Italian geometer, and a later disciple of Galilei. He is best remembered by his work *Dissertation of the Fifth Book of the Conic Sections of Apollonius*, which contained a multitude of new properties of these curves. He was a member of the Accademia del Cimento.

WARD, SETH (1616–88).—An eminent Divine. He graduated M.A. at Sidney Sussex College, Cambridge, in 1640, and became Fellow of his college in the same year. He received his mathematical instruction from Oughtred, and he became a lecturer in that subject in 1643. In the same year he wrote against the Solemn League and Covenant, and was ejected from his Fellowship. From 1649 to 1661 he held the Savilian Professorship of Astronomy, and he acquired a high reputation by his Theory of Planetary Motion. He became Principal of Jesus College, but he was ejected by Cromwell in 1657.

He became D.D. in 1654, and after that his promotion in the Church was rapid. He became Bishop of Exeter in 1662, whence he was translated to Salisbury in 1667. He was particularly severe against Dissenters, and he was associated with Wallis in his quarrel with Hobbes. Aubrey describes him as the " pattern of humility and courtesie " *, but Wood is very severe against him. " The said Doctor Ward " he says, " did about his Majesties Restauration 1660 endeavour to make his loyalty known by being imprisoned at Camb., by his ejection, by his writing against the Covenant, and I know not what, but not a word of his cowardly wavering for lucre and honour sake, of his putting in and out, and occupying other men's places for several years " †.

He published many mathematical and theological treatises.

WILKINS, JOHN (1614–72).—Was one of the founders of the Royal Society, and at its first meeting in December 1660 he took the Chair. He studied at Oxford (Magdalen Hall), where he became M.A. in 1643. He subscribed to the Solemn League and Covenant, yet he was equally prepared to swear allegiance to Charles II on his Restoration. In 1656 he married the sister of Oliver Cromwell, and by his son, Richard Cromwell, he was made Master of Trinity College in 1659. Nine years later he was raised to the See of Chester. His first publication was *The Discovery of a New World* (1638). All his mathematical and philosophical works were collected and published in 1708.

Aubrey describes him as a " very ingeniose man ; he had a very mechanicall head. He was much for trying of experiments, and his head ran much on perpetual motion ". Evelyn speaks of him as a " most obliging person, who took great pains to preserve the Universities from ignorant and sacrilegious commanders and soldiers " ‡. He went on to say that " he has in his lodgings a variety of shadows, dials and many other artificial mathematical and magical curiosities ". About 1649 he went to Oxford, where, according to Aubrey, " he was the principal reviver of experimental philosophy (secundum mentem Domini Baconi) at Oxford, where he had weekely, an experimental philosophicall clubbe . . . . which began in 1649 and which was the incunabile of the Royall Society " §. Wallis was several times President of this Society.

DE WITT, JAN (1625–72).—A statesman of Holland who cultivated the new Cartesian methods in geometry. His work has been preserved for us in the Commentaries of Van Schooten. His most valuable treatise is his *Elementa Linearum Curvarum* (Amsterdam, 1658). In this he conceived a new and ingenious way of generating Conics which is essentially the same as that by projective pencils of rays in modern synthetic geometry. He treated his subject not synthetically but, like Wallis, with the aid of the Cartesian Analysis.

* Aubrey, ii, p. 576.
† Wood, *Athen. Oxon*, ii, 686.
‡ Evelyn, p. 228.
§ Aubrey, *Lives*, ii, 583.

WILKINSON, HENRY (1616–90).—He was Principal of Magdalen College, Oxford. He was also one of the Parliamentary Visitors of Oxford in 1647. He was ejected from his position at Magdalen Hall for political reasons in 1662 (Wood, *Ath. Oxon*, iv 285).

WOOD, ANTHONY À (1632–95).—This noted antiquarian and historian was educated at New College, where he received his M.A. in 1655. In 1674 he published his *Historia et Antiquitates Univ. Oxon*. His *Athenæ Oxonienses*, " an exact History of all the Writers, and Bishops who have had their education in Oxford, from 1500 to 1690 ", was published in 1690–91. This work excited very bitter feelings on account of its reckless charges and bitter criticisms, one of which led to his prosecution by the young Earl of Clarendon in 1692, for libelling his father. For this he was condemned, and expelled from the University.
Several MSS. of his were published posthumously. For many of his works he received great assistance from Aubrey and others, but this was very imperfectly acknowledged.

WREN, Sir CHRISTOPHER (1632–1723).—This famous architect was also a distinguished mathematician and astronomer. He was Fellow of All Souls College from 1653 to 1661, Professor of Astronomy at Gresham College, 1657 to 1661, Savilian Professor of Astronomy, 1661 to 1673, D.C.L. Oxford and LL.D. Cambridge. He was elected to a Fellowship of the Royal Society in 1663.
Whilst at Oxford, Wren distinguished himself as a mathematician, and Newton, in his *Principia* (p. 19, 1713 edition), spoke very highly of his work as such.
Wren initiated a number of experiments on the subject of the variation in the height of the barometer. He also devoted much time and attention to medical and anatomical subjects. He was a prominent member of the circle which ultimately became the Royal Society, and from 1680 to 1682 he was its President. During this period he made contributions on a number of subjects, particularly the Impact of Bodies. He probably applied himself to architecture about 1663. In that year he built the Chapel of Pembroke College and the next year the Sheldonian Theatre. He prepared a scheme for the rebuilding of London after the Great Fire in 1666. He is said to have built no less than 52 London Churches, but it seems that not a few of these were due to Hooke. He never printed anything himself, but several of his works have been published by others, and there are numerous papers of his in the *Transactions*.

ZOUCH, RICHARD (1590–1660).—Was educated at Winchester College and New College, Oxford. He became a Fellow of the latter (1609), and later was made Warden of the Cinque Ports. " In 1648 when the Visitors appointed by Parliament sate in the University, he submitted to their power, and so kept his Principality and professorship during times of the usurpation. He was an exact Artist, a subtile Logician, expert Historian, and for the Knowledge in and practice of the Civil Law the chief Person of his time, and his works much esteemed beyond the Seas. He was so well vers'd also in the Statutes of the University and controversies between the members thereof and the City that none after *Twynes'* death went beyond him. As his birth was noble, so was his behavior and discourse " *. He left many legal works.

* Wood, *Athen. Oxon,* ii, col. 166.

APPENDIX II.

Some Observations on the Development of Notation during the Seventeenth Century, with Specimens of Notation then Current.

Not the least service which Wallis rendered to the development of mathematics during the seventeenth century was the impulse he gave to a movement which had already begun to take shape towards the end of the previous century, namely, the transition from the mathematical symbolism employed by earlier writers to the modern language of mathematics. The progress of algebra in particular had been seriously hampered during the early decades of the century by the crudeness of the notation which still seemed to find favour among even distinguished writers. The development of a clearly-defined system proceeded with discouraging slowness. In the writings of mathematicians from say, 1560 onwards, we recognise two distinct tendencies. In the first, the rhetorical style still persisted; no better example of this, perhaps, can be cited than Cavalieri's *Geometria Indivisibilibus* (1635), which is as rhetorical in its exposition as is the original text of Euclid's Elements*. No use whatever is made of the arithmetical or algebraical signs; not even do we meet p or m, still less $+$ or $-$, and all the operations are described in words. Even one hundred years after Recorde's *Whetstone of Witte* there were many writers who did not use any symbol for equality. In the other case, we note a decided inclination to go to the other extreme. Not only were symbols employed for the usual operations of addition, subtraction, etc., new symbols were invented to express the sum, or difference, etc., of various quantities. This movement reached its climax in the writings of Oughtred, a specimen of whose work is appended.

Wallis was one of the first who tried to get the best out of two worlds. He surveyed the medley of notations current in his day, and he made a deliberate attempt to evolve from them a system which could be easily understood. He abandoned the excesses of Oughtred, and from the conflicting elements he derived a system not widely different from our own. Unfortunately, even he vacillated. In the *Arithmetica Infinitorum* he used the Index Notation for powers throughout. Thirty years later in the *Algebra* he fell back on the clumsy notation ($bbbb$ for b^4) adopted by Harriot. A distinctly retrograde step, especially as in the case where the powers came to be numerous, he did not always exercise sufficient care, so that frequently we find $bbbbb$ for b^6. The establishment of a well-defined system for radicals seemed to be equally remote. Very great diversity prevailed about this time as to the exact position of the numeral relative to the root in the radical sign. A step in the right direction was taken by Wallis (*Arithmetica Infinitorum*) when he expressed the root indices in numerals without enclosing them in a circle, as did

---

* Cajori (*Notations*), p. 186.

APPENDIX II

earlier writers (*e. g.*, Stevinus), or in parentheses. His method of writing $\sqrt{3}R^2$ for $\sqrt[3]{}R^2$ was a distinct improvement. But his system for radicals was not invariable; even on the same page we meet $\sqrt{}qqR$ for $\sqrt[4]{}R$ and $\sqrt{}^6R$ for $\sqrt[6]{}R$.

Wallis was very anxious to see algebraic notation placed on a clear foundation. His views are well expressed in the following passage, which is taken from his *Algebra* (p. 68):—

" I find it also of very good use many times, (in designing Quantities by Symbols, Species, or Notes to be taken at pleasure) to make choice of such Notes or Species as may some way represent to the Memory or Fancy the Quantities designed by them.

" For though such choice of Notes do not at all influence the Demonstration, yet it doth assist the Fancy and Memory, which would otherwise be in danger of being confounded in a Multitude of Symbols; especially if in each several Proposition, the same Notes or Symbols come to signify different things.

" And without this advantage, it would have been impossible, (for instance) in my Prop. 13–22. Cap. V. *De Motu*, to have managed those perplexed Computations, without great prolixity and confusion, if I had not, for assisting the Memory, made choice of suitable Symbols for each several Quantity, and kept constant to them through the whole Discourse . . . .

" Such advantages as these, (and others as the occasion of the subject may require) I oft find very useful, (especially where the Symbols be numerous) for assisting the Fancy, and easing the Memory, and bringing the whole Process to as narrow a prospect as may be; which thereby becomes intelligible, with much more ease than when involved in a multitude of Words, and long Periphrases of the several Quantities and Operations ".

In the following pages, specimens of notation, all of them current during the seventeenth century, are exhibited. It will be noticed at once that the notation employed in the *Conic Sections* is not unlike that which one would expect to find in a modern treatise. One cannot fail to notice the use of a^2 to express the second power of a, which Descartes and even later writers have written aa.

1. OUGHTRED : *Clavis Mathematicæ*, 1631.

   Note :

   (1) The introduction of new symbols to indicate
   - (*a*) the sum (or difference) of two quantities ;
   - (*b*) the sum (or difference) of their squares ;
   - (*c*) the square of their sum (or difference) ;
   - (*d*) the sum (or difference) of their cubes, etc.

   (2) The sign $\sqrt{}$b for *binomial root*.

   (3) Method of writing " quantities proportional ".

Clavis Mathematicæ

CAP. XI.

Exempla aliquot facillima, quibus quæ hactenus tradita sunt familiaria redduntur: Et via ad Æquationem Analyticam sternitur.

Sciendum primò est, quod in sequentibus, tum brevitatis, tum phantasiæ juvandæ gratia, passim ferè his verborum symbolis utor. A & E significant duos numeros, sive magnitudines; quorum A plerumque major est, E minor. Æ rectangulum sub ipsis. Z est summa. X differentia. Zq summæ quadratum. Xq differentiæ quadratum. Z summa quadratorum. X differentia quadratorum. Z summa cuborum. X differentia cuborum. A, M, E, sint tres continuè proportionales: A, M, N, E, quatuor. Q: C: QQ: QC: &c. præfixæ magnitudinibus inter duo utrinq; puncta inclusis, significant illiusmodi potestates. √ denotat radicem sive latus potestatis simplicis, si non intercedant duo puncta: Si vero potestas duobus utrinque punctis includatur, significat latus ipsius universale: quod etiam aliter per literam b vel r describi solet, ut √b latus est Binomii, & √r latus Residui sive Apotomes. = nota est æqualitatis.

2. Sunt duo numeri sive magnitudines, quorum major est A, minor E: quænam est ipsorum summa? quæ differentia? quod sub ipsis rectangulum? quæ quadratorum summa? quæ quadratorum differentia? quæ summæ & differentiæ ipsorum summa? quæ summæ & differentiæ ipsorum differentia? quod summæ & differentiæ ipsorum rectangulum? quod summæ quadratum? quod differentiæ quadratum? quæ quadratorum summæ & differentiæ summa? quæ quadratorum summæ & differentiæ differentia? quod quadratum rectanguli?

Z est $A+E$. X est $A-E$. $Æ$ est AE.
$Z = Aq+Eq$.
$Z+X = 2A$. $Z-X = 2E$.
$\frac{1}{2}Z+\frac{1}{2}X = A$. $\frac{1}{2}Z-\frac{1}{2}X = E$.
$ZX = Aq-Eq = X$. $Zq.X :: Z.X$.
$Zq = Aq+2AE+Eq = Z+2Æ$.
$Xq = Aq-2AE+Eq = Z-2Æ$.
$Zq+Xq = 2Aq+2Eq = 2Z$.
$Zq-Xq = 4AE$. $\frac{1}{2}Zq-\frac{1}{2}Xq = Æ$.
$Æq = AqEq$.

3. Sunt duo numeri sive magnitudines, quorum summa est Z, & major ex ipsis ponitur A: quisnam est minor? quæ ipsorum differentia? quod sub ipsis rectangulum? quæ quadratorum summa? quæ quadratorum differentia?

$E = Z-A$. $X = 2A-Z$. $Æ = ZA-Aq$.
$Z = Zq-2ZA+2Aq$. $X = 2ZA-Zq$.

Si vero minor ex ipsis ponatur E:
$A = Z-E$. $X = Z-2E$. $Æ = ZE-Eq$.
$Z = Zq-2ZE+2Eq$. $X = Zq-2ZE$.

4. Sunt duo numeri sive magnitudines, quorum differentia est X, & major ex ipsis ponitur A: quisnam

2. HARRIOT : *Artis Analyticæ Praxis.*

   Note :

   (1) Method of writing powers.

   (2) Full stop between coefficient and literal quantities.

   (3) Method of writing cube root $\sqrt{3}$.

APPENDIX II
SECTIO SEXTA.

$$9 \equiv\!=\!\equiv -6a + aaa \ldots a \equiv\!=\!\equiv \sqrt{3)\tfrac{9}{2}} + \sqrt{\tfrac{81}{4}} + \sqrt{3)\tfrac{9}{2}}$$
$$- \sqrt{\tfrac{81}{4}} \equiv\!=\!\equiv 3.$$

Nota 1.

Æquationem istam $aaa - 3.bba \equiv\!=\!\equiv + 2ccc$. propter similitudinem quæ inter tres illius casus, & sectiones conicas hyberbolem, parabolam, & ellipsim, in triplici differentia excessus, æqualitatis & defectus intercedit, similibus nominibus, hyberbolicam scilicet, parabolicam & ellipticam appellare licet. Hyperbolicam in qua c. maior est quam b. parabolicam in qua c. ipsi b. æqualis est, ellipticam in qua c. minor est quam b. atque eam ob causam (in specie) irresolubilem.

Nota 2.

In duabus antecedentibus æquationibus accidit interdum binomia cubica solutionis radicalibus implicata explicari posse per radices itidem binomias, quæ per summam vel differentiam constituant tandem radicem simplicem æquationis explicatoriam. Huius generis solutionum exempla sunt quæ sequuntur.

$$52 \equiv\!=\!\equiv -3a + aaa \ldots a \equiv\!=\!\equiv 4.$$
$$a \equiv\!=\!\equiv \underbrace{\sqrt{3.)26 + \sqrt{675}}} + \underbrace{\sqrt{3.)26 - \sqrt{675}}}.$$
$$\underbrace{2. + \sqrt{3}} \ldots \underbrace{+ \ldots 2 - \sqrt{3}}.$$
$$4.$$

$$270 \equiv\!=\!\equiv +9a + aaa \ldots a \equiv\!=\!\equiv 6.$$
$$a \equiv\!=\!\equiv \underbrace{\sqrt{3.)\sqrt{18252} + 135}} - \underbrace{\sqrt{3)\sqrt{18252} - 135}}.$$
$$\underbrace{\sqrt{12} + 3} \ldots \underbrace{- \ldots \sqrt{12} - 3}.$$
$$6.$$

$$40 \equiv\!=\!\equiv -6a + aaa \ldots a \equiv\!=\!\equiv 4.$$
$$a \equiv\!=\!\equiv \underbrace{\sqrt{3)20 + \sqrt{.392}}} + \underbrace{\sqrt{3.)20 - \sqrt{'292}}}.$$
$$\underbrace{2. + \sqrt{.2.}} \underbrace{+ \ldots 2 - \sqrt{.2.}}.$$
$$4.$$

$$20 \equiv\!=\!\equiv +6a + aaa \ldots a \equiv\!=\!\equiv 2.$$
$$a \equiv\!=\!\equiv \underbrace{\sqrt{3.)\sqrt{.108} + 10}} - \underbrace{\sqrt{3.)\sqrt{108} - 10}}.$$
$$\underbrace{\sqrt{.3} + 1} \ldots \underbrace{- \ldots \sqrt{.3} - 1}.$$
$$2.$$

3. WALLIS : *De Sectionibus Conicis* (1655).

   Note :

       (1) Use of index notation, even to squares.

       (2) Use of dots : : after root sign to indicate the root of the whole expression.

APPENDIX II

PROP. 35. *De Sectionibus Conicis.* 83

trum-transversam terminantes, *Vertices Oppositi*: Punctumq; inter oppositos vertices medium, est utriq; Hyperbolæ centrum commune.

Hæ autem Hyperbo'æ Oppositæ, respondent Hyperbolis Oppositorum conorum eodem plano sectorum.

PROP. XXXV.

Corollaria.

Patet ex dictis; Cùm sit $h^2 = ld + \frac{l}{t} d$: Horum in Hyperbola quatuor, *Diametro-transversâ, Diametro-interceptâ, Latere-recto, & Ordinatim-applicatâ*; Datis tribus quibusvis, etiam reliquum (magnitudine) dari. Nempe (disponendo & resolvendo æquationes) erit

In Hyper-bola,
$$\begin{cases} \text{Ordinatim-applicata,} \ h = \sqrt{: ld + \frac{l}{t} d^2 :} = \sqrt{\frac{td + d^2}{t} l}. \\ \text{Ejusq; quadratum,} \ h^2 = ld + \frac{l}{t} d^2 = \frac{td + d^2}{t} l. \\ \text{Latus-Rectum,} \ l = \frac{h^2}{td + d^2} t. \\ \text{Diameter-transversa,} \ t = \frac{td + d^2}{l} l. \\ \text{Diameter-intercepta,} \ d = \sqrt{: \tfrac{1}{4}t^2 + \frac{t}{l} h^2 :} - \tfrac{1}{2}t. \end{cases}$$

Adeoq; ; Diametrorum transversæ & interceptæ aggregatum, $t + d = \sqrt{: \tfrac{1}{4}t^2 + \frac{t}{l} h^2 :} + \tfrac{1}{2}t.$

Distantia puncti applicationis a centro. $c = \sqrt{: \tfrac{1}{4}t^2 + \frac{t}{l} h^2}.$

Patet

4. WALLIS : *Arithmetica Infinitorum*, 1656.

   Note :

   (1) Use of index notation. (In this illustration he is approaching the idea of negative exponents.)

   (2) Symbol for proportion : :

Arithmetica Infinitorum. Prop. 88.

es dividens est uno gradu superior serie dividenda, adeoque index seriei dividentis unitate major quam index seriei divisæ, puta $3-2=2-1=1-0=1$:) termini seriei oriundæ erunt reciproce-proportionales homologis terminis seriei primanorum. Puta si respective dividatur

series $0a^2$, $1a^2$, $4a^2$, $9a^2$, $16a^2$,   &c.
per seriem $0a^3$, $1a^3$, $8a^3$, $27a^3$, $64a^3$,   &c.

vel series $0a$, $1a$, $2a$, $3a$, $4a$,   &c.
per seriem $0a^2$, $1a^2$, $4a^2$, $9a^2$, $16a^2$, &c.

vel series 1, 1, 1, 1, 1,   &c.
per seriem $0a$, $1a$, $2a$, $3a$, $4a$, &c.

prodibit series $\frac{1}{0a}$, $\frac{1}{1a}$, $\frac{1}{2a}$, $\frac{1}{3a}$, $\frac{1}{4a}$, &c.

cujus seriei termini sunt reciproce proportionales homologis terminis seriei primanorum $\frac{0a}{1}$, $\frac{1a}{1}$, $\frac{2a}{1}$, $\frac{3a}{1}$, $\frac{4a}{1}$, &c.

ut patet. Nempe $\frac{1}{2a}$. $\frac{1}{3a}$. :: $\frac{3a}{1}$. $\frac{2a}{1}$. Et sic ubique.

Eodem modo si series primanorum dividenda sit per seriem tertianorum; vel (quod tantundem valet) series Æqualium per seriem Secundanorum; erit series proveniens seriei secundanorum reciproce proportionalis. puta

$\frac{1}{0a}$, $\frac{1}{1a}$, $\frac{1}{4a}$, $\frac{1}{9a}$, $\frac{1}{16a}$, &c.

Et pari modo in omnibus ejusmodi divisionibus continget.

PROP. LXXXVIII. *Coroll.*

SI infinita plana (*parallela*) Parallelepipedi, ad totidem rectas Trianguli (*æquè-alti*) applicentur; (*vel si respectivis rectis Trianguli, & Parallelogrammi, sumantur tertiæ proportionales*;) series rectarum provenientium erit reciproca seriei Primanorum, quæ quidem

rectæ

5. WALLIS : *Algebra*, 1685.

Note reversion to Harriot's method of writing powers.

APPENDIX II

Chap LI. *Of Biquadratick Equations.*

Becomes $\quad eeee - 6bbee + 8bbbe = -cccc + 3bbbb.$

Whose Root is $\quad e = a + b.$

XVIII. The Equation, $\quad aaaa + 4baaa + ddaa = +cccc;$

Putting $\quad a = e - b;$

And consequently,

$$\left.\begin{array}{l} eeee - 4beee + 6bbee - 4bbbe + bbbb = + aaaa \\ +4beee - 12bbee + 12bbbe - 4bbbb = +4baaa \\ \quad\quad\quad + ddee - bddd = + dda \end{array}\right\} = +cccc;$$

Becomes $\quad\begin{array}{l} eeee - 6bbee + 8bbbe = + cccc \\ \quad + ddee \quad\quad + 3bbbb \\ \quad\quad\quad\quad + bddd. \end{array}$

Whose Root is $\quad e = a + b.$

XIX. The Equation, $\quad aaaa - 4baaa - ddaa = +cccc;$

Putting $\quad a = -e + b;$

And consequently,

$$\left.\begin{array}{l} eeee - 4beee + 6bbee - 4bbbe + bbbb = + aaaa \\ +4beee - 12bbee + 12bbbe - 4bbbb = -4baaa \\ \quad\quad\quad + ddee - bddd = - dda \end{array}\right\} = +cccc;$$

Becomes $\quad\begin{array}{l} eeee - 6bbee + 8bbbe = + cccc \\ \quad + ddee \quad\quad + 3bbbb \\ \quad\quad\quad\quad + bddd. \end{array}$

Whose Root is $\quad e = -a + b.$

Or putting $\quad a = +e + b;$

And consequently,

$$\left.\begin{array}{l} eeee + 4beee + 6bbee + 4bbbe + bbbb = + aaaa \\ -4beee - 12bbee - 12bbbe - 4bbbb = -4baaa \\ \quad\quad\quad - ddee - bddd = - dda \end{array}\right\} = +cccc;$$

Becomes $\quad\begin{array}{l} eeee - 6bbee - 8bbbe = + cccc \\ \quad - ddee \quad\quad + 3bbbb \\ \quad\quad\quad\quad + bddd. \end{array}$

Whose Root is $\quad e = +a - b.$

XX. The Equation, $\quad aaaa + 4baaa + ffaa = +cccc;$

Putting $\quad a = e - b;$

And

APPENDIX III.

List of Wallis's Mathematical Works, including his Contributions to the *Transactions*.

Wallis's mathematical and scientific works were collected and published together in the *Opera* (Three Volumes Folio), 1693–99. The following is a *List of Contents* :—

VOLUME I.

Oratio Inauguralis.
Mathesis Universalis, seu Opus Arithmeticum.
Adversus M. Meibomii *De Proportionibus Dialogum*.
De Sectionibus Conicis. Nova Methodo Expositis.
Arithmetica Infinitorum.
Eclipsis Solaris Observatio, Oxonii habita 2 Aug. 1654.
De Cycloide Tractatus.
Tractatus Epistolaris, ad D. Hugenium.
Mechanica ; sive De Motu, Tractatus Geometricus.

VOLUME II.

De Algebra Tractatus, Historicus et Practicus.
    (This had already been published in English in 1685.)
De Combinationibus, Alternationibus, et Partibus Aliquotis, Tractatus.
De Sectionibus Angularibus, Tractatus.
De Angulo Contactus et Semicirculi, Tractatus.
Ejusdem Tractatus Defensio.
De Postulato Quinto, et Quinta Definitione Lib. vi Euclidis; Disceptatio Geometrica.
Cono-Cuneus.
De Gravitate et Gravitatione. Disquisitio Geometrica.
De Æstu Maris. Hypothesis Nova.
Commercium Epistolicum. De Quæstionibus quibusdam Mathematicis habitum.

VOLUME III.

Claudii Ptolemæi. Harmonicorum Libri Tres.
Porphyrii. In Harmonica *Ptolemæi* Commentarius.
Manuelis Bryennii. Harmonica.
Archimedis. Arenarius et Dimensio Circuli.
Aristarchi Samii : De Magnitudinibus et Distantiis Solis et Lunæ.
Pappi Alexandrini, Libri Secundi Collectionum Mathematicarum (hactenus desiderati) Fragmentum.
Epistolarum quarundam *Collectio* Rem Mathematicam spectantium.
Opuscula quædam Miscellanea.

APPENDIX III

In addition to the above-published works, Wallis contributed a number of papers to the *Philosophical Transactions*.

These were :—

1. Observations of the sealed Weather glass, and the Barometer.   I. 163.
2. A Relation concerning the late Earthquake near Oxford, Jan. 19, 1665.   I. 166.
3. A Relation of an Accident by Thunder and Lightning at Oxford, May 10, 1666.   I. 222.
4. An Essay exhibiting his Hypothesis about the Flux and the Reflux of the Sea.   I. 263.
5. An Appendix by way of an Answer to some Objections to the above Essay.   I. 281.
6. Animadversions of Dr. Wallis upon Master Hobs's Book : *De Principiis et Ratiocinatione Geometrarum*.   I. 289.
7. Some Inquiries and Directions concerning Tides, proposed by Dr. Wallis.   I. 297.
8. Account of the Variety of the Annual High-Tydes as to several places, with respect to his own Hypothesis.   III. 652.
9. Some Mistakes to be found in a Book under the title of *Specimina Mathematica*, by F. du Laurens.   III. 654.
10. Animadversions on a Printed Paper entitul'd : Responsio F. du Laurens ad Epist. D. Wallisii ad Oldenburgium scriptam.   III. 744.
11. Second Letter on the same Paper.   III. 747.
12. Continuation of the Second Letter.   III. 825.
13. Account of the *Logarithmotechnia* of Mercator.   III. 753.
14. A Summary Account of the General Laws of Motion.   III. 864.
15. Some Observations concerning the Baroscope and the Thermoscope.   IV. 1113.
16. Letter to Boyle, concerning an Essay of Teaching a Person Deaf and Dumb to speak, with the Success of it.   V. 1087.
17. Answer to Mr. Childrey's Animadversions upon his Hypothesis of the Flux and the Reflux of the Sea.   V. 2068.
18. Answer to Mr. Hobbes' *Rosetum Geometricum*.   VI. 2202.
19. His Opinion concerning the Hypothesis *Physica Nova* of Leibniz.   VI. 2227.
20. Answer to Four Papers of Mr. Hobs, lately published.   VI. 2241.
21. Breviat concerning Dr. Wallis's two Methods of Tangents.   VII. 4010.
22. An Answer to the *Lux Mathematica*.   VII. 5067.
23. On the Centre of Gravity of the Hyperbola.   VII. 5074.
24. On the Suspension of Quicksilver, well purged of Air, much higher than the ordinary Standard in the Torricellian Experiment.   VII. 5160.
25. A Confirmation of the Account of a Strange Freezing in Somersetshire.   VIII. 5196.
26. A Note upon Mr. Lister's Observation concerning Veins in Plants.   VIII. 6060.

27. Letter asserting the first Invention and Demonstration of the Equality of the curve line of a Paraboloeid to a Straight Line, and next, the finding a Straight Line equal to that of a Cycloid, and of the Parts thereof. VIII. 6146.
28. Two other Letters to the same Purpose. VIII. 6149.
29. A Letter gratulatory to M. Hevelius for his *Organographia*. IX. 243.
30. On a new musical Discovery. XII. 839.
31. An Account of a considerable Meteor, seen in many Distant Places of England, Sept. 20, 1676. XII. 863.
32. An Account of an Antient Date in Northamptonshire in Numeral Figures. XIII. 399.
33. An Account of two large stone Chimney Pieces, with a peculiar Sort of Archwork thereon. XIV. 800.
34. On the Collection of Secants, and the True Division of the Meridian in the Sea Chart. XV. 1193.
35. On the Air's Gravity. XV. 1002.
36. Treatise of Algebra, Historical and Practical. XV. 1095.
37. An Account of the Strength of Memory, when applied with due attention. XV. 1269.
38. On the Measures of the Air's Resistance to Bodies moved in it. XVI. 269.
39. On the apparent Magnitude of the Sun and the Moon, or the apparent Distances of two Stars when near the Horizon, and when higher elevated. XVI. 323.
40. The Florentine Problem, concerning the Quadrature of a hemispherical Cupola of a Temple at Delos. XVII. 584.
41. A Proposal concerning the Parallax of the Fixed Stars, in reference to the Earth's Annual Orb. XVII. 844.
42. A Discourse concerning the Methods of Approximation in the Extraction of Surd Roots. XIX. 2.
43. On the Spaces in the Cycloid, which are perfectly Quadrable. XIX. 111.
44. An extraordinary Cure of a Horse that was staked into his Stomach. XIX. 178.
45. Letter concerning the Cycloid known to Cardinal Cusanus about 1450, and to Carolus Borillus about 1500. XIX. 561.
46. On the Generation of Hail and Thunder and Lightning, and the Effects thereof. XIX. 729.
47. A Correction of the 109th Chapter of his book on *Algebra*. XIX. 729.
48. An Account of the Effects of Thunder and Lightning at Everdon, Northamptonshire. XX. 5.
49. A Question in Musick lately proposed to Dr. Wallis concerning the Divisions of the Monochord, or Section of the Musical Canon, with his Answer to it. XX. 80.
50. On the Observation of Easter, April 24, 1698. XX. 185.
51. On the supposed Imperfections in an Organ. XX. 249.
52. On the strange Effects of Musick in former times. XX. 297.

APPENDIX III

53. A Method of instructing Deaf and Dumb Persons. XX. 353.
54. An Account of some Passages between him and M. Leibniz. XXI. 273.
55. Letter to Leibniz. XXI. 280.
56. On the supposed Alteration of the Meridian Line, which may affect the Declination of the Magnetic Needle, and the Pole's Elevation. XXI. 285.
57. On the Alteration (suggested) of the Julian Calendar for the Gregorian. XXI. 343.
58. The Quadrature of the Parts of the Lunula of Hippocrates, performed by Mr. John Perks, with the further improvements of the same by David Gregory and John Caswell. XXI. 411.
59. Some easy Methods for the measuring of Curved-lined Figures, plain and solid. XXII. 547.
60. On the Use of Numeral Figures in England in 1090. XXII. 677.
61. Two Letters on Men feeding on Flesh. XXII. 783.
62. A Brief Relation of some Strange Bones lately digged up in some grounds of Mr. Joh. Sommers in Canterbury. XXII. 882.
63. Some Letters relating to Mr. Sommer's Treatise on Chartham News. XXII. 1022.
64. Letter relating to that Isthmus which is supposed to have joined England to France in former times, where now is the Passage between Dover and Calais. XXII. 967.
65. Captain Edm. Halley's Map of Magnetic Variation, and some other things relating to the Magnet. XXIII. 1106.

Note: The above Papers cover the Period from 1665 to 1702.

*[Paper No. 62 is not by Wallis.—ED.]

BIBLIOGRAPHY.

A. LIST OF MANUSCRIPTS CONSULTED.

The Journal Book of the Royal Society.
The Letter Book of the Royal Society.
The Register of the Royal Society.
Wallis's Letters, mainly to different members of the Society.
Letters from Members of the Society to Wallis.
The Savile MS. (Bod.).
The Smith MSS. (Bod.).
The Tanner Papers (Bod.).
Add. MSS. (Bod.).
MS. Life of Wallis (British Museum).   Addit. MSS. 32601 (1735).
Rigaud MSS. (Bod.).

B. In addition to the works mentioned in the Preface (pp. vi–ix), to which specific reference is made in the text, the following have also been consulted :—

Dampier-Whetham, W. C. D. A History of Science and its Relation with Philosophy and Religion. Camb., 1930.
De Morgan. A Budget of Paradoxes. Lond., 1872.
Duhem, Pierre. Les Origines de la Statique.
Greenstreet, W. J. Isaac Newton, 1642–1727. Lond., 1927.
Gunther, R. T. Early Science in Oxford. (Oxf. Hist. Soc.)
Lagrange, J. L. Mécanique Analytique. Paris, 1788.
MacPike, E. F. Correspondence and Papers of Edmond Halley. Lond., 1932.
Sedgwick and Tyler. A Short History of Science. New York, 1929.
Sullivan, J. W. N. The History of Mathematics in Europe. Lond., 1925.
Tanner, Jos. Robson. Private Correspondence and Miscellaneous Papers of Samuel Pepys (1679–1703). Lond., 1926.
Waller, Richard. Posthumous Works of Robt. Hooke. Lond., 1705.
Wolf, A. A History of Science, Technology, and Philosophy in the XVIth and XVIIIth Centuries. Lond., 1934.

INDEX.

(The figures in heavy type refer to the List of Names in Appendix I. The letter " n " following a number indicates a footnote.)

Academia Secretorum Naturæ, 8 n.
Académie des Sciences, 176, 185, 190, 202, 204, 209, 210.
Accademia dei Lincei, 8 n.
Accademia del Cimento, 184, 189, 214.
Algebra, Treatise of, 133.
Ancient Manuscripts, Publication of, 130.
Anderson, Alexander, 149, **180**.
Angle of Contact, 65.
    ,,   ,,   *Defense of*, 66, 164.
Apollonius, 17, 18, 21, 22, 23.
Archimedes, 26, 77, 91.
Areas greater than Infinite, 43–46.
Aristotle, 2, 91, 92, 93.
Arithmetical Progressions, 70, 71.
Artis Analyticæ Praxis, 138, 144, 145.
Aubrey, John, 83, 187, 192, 199, 208, 209, 214, 215, 216, **180**.

Bacon, Francis, 2, 4, 7.
Baillet, A., 150, 155.
Baker, Thomas, 163, **181**.
Baliani, 101, 108, 110, **181**.
Ball, W. R. R., 133.
Barrow, Isaac, 5, 162, 187, 190, 210, **181**.
Bartholinus, 162, **182**.
de Beaune, 97.
Benedetti, 93, 94, **183**.
Bernouilli, 74.
Berkeley, 172.
Binomial Theorem, 61.
Birch, Thomas, 152.
Blencow, William, 177.
Bombelli, 134, 137, 161, **183**.
Borelli, 98, 189, **183**.
Boyle, Robert, 2, 9, 11, 87, 89, 99, 153, 167–8, 205, 206, 207, 208.
Branker, Thomas, 187, 208.
Briggs, Henry, 5, **184**.
Brouncker, Lord, 60, 62, 64, 71, 72, 74, 75, 81, 82, 107, 179, 190, 193, **184**.
Byron, Sir John, 14.
Burgess, Anthony, 4.

INDEX

Byfield, Adoniram, 6, 83, **185.**

Calendar Reform, 131.
Carcavi, 154, **185.**
Cardan, 91, 92, 140, 142, 158, 159, 162, **185.**
Castelli, 95, 186, 213, **186.**
Caswell, John, 165, **186.**
Cavalieri, 15, 18, 19, 20, 26, 27, 28, 151, 209, 218, **186.**
Centre of Gravity, 114.
Centre of Percussion, 120.
Characteres cossici, 67.
Chuquet, 35.
Clavis Mathematicæ, 61, 138, 146, 147.
Clavius, 65, 66, 68, 70, 210, **187.**
Collins, J., 1, 13, 75, 81, 87, 129,136, 137, 147, 148, 152, 155, 162, 183, 188, 190, 193, 203, 204, **187.**
Combinations, Alternations and Aliquot Parts, 164.
Commercium Epistolicum (1657/8), 75, 184.
    ,,   ,,   ,, *(1712)*, 187, 194.
Conic Sections, 186, 205, 210, 211, 213, 214, 215.
Continued Fractions, 60, 184.
Cotes, Roger, 156, **187.**
Cubic Equation, 183, 185, 191.
Cuno-cuneus, 87.
Curll, Dr. Walter, 6, **188.**
Cusa, Cardinal, 204.
Cycloid, Prize Questions, 84, 203.

Darley, Sir Richard, 6.
Decimals, 69, 184, 206, 212.
De Cissoide, 84.
De Cycloide, 80–82, 84.
De Morgan, A., 133, 134.
De Motu, 84, 90.
Descartes, R., 2, 5, 15, 22, 23, 24, 67, 68, 73, 96, 99–105, 109, 133–140, 148–151, 155, 158, 160–3, 165, 183, 188, 190–192, 199, 201, 204–5, 209, 211.
Descartes' Rule of Signs, 158.
" Dettonville, A.", 154.
Differential Calculus, 131.
Digby, Sir Kenelm, 71, 72–76, 78 n, 80, **188.**
Du Laurens, F., 88, 148, 153, 155, **188.**

Elasticity, 104.
Ellipse, 22.
Energy, 112.
Ent, Sir George, 8, **189.**

INDEX 237

" Equipollents ", 139.
Eudoxus, 26.
Euler, 74, 79.
Evelyn, John, 11, 13, 215.

Fabri, 98, **189.**
False Roots, 136.
Fell, John, 177, **189.**
Fermat, 2, 5, 15, 27, 28, 64, 71–82, 84, 88, 118, 135, 151, 152, 161, 163–4, 190, 201, 204.
Figurate Numbers, 51–55.
Flamsteed, 128, 129, 173, 193, **190.**
Fluxions, Method of, 19, 66.
Foster, Samuel, 8, 10, **190.**
Force, 108, 109.
Fractional Indices, 34–38.
Frénicle de Bessy, 72–75, 155, **190.**
" Fuga Vacui ", 102.

Galilei, Galileo, 2, 89, 93, 95, 96, 98–101, 109, 112–114, 124, 183, 186, 196, 199, 202, 204, 213, 214.
Gassendi, 199, **191.**
General Methods, 25, 34, 79, 112.
" Geometrical Effections ", 161.
Geometrical Progressions, 70.
Getaldus, 161.
Gilbert, William, 2.
Girard, Albert, 93, 150, 151, 159, 160, **191.**
Glisson, Francis, 4, 8, **192.**
Goddard, Jonathan, 8, 9, 10, 66, **192.**
Gregorian Calendar, 131–2.
Gregory, David, 64, 84, 106, 170, **193.**
   ,,   James, 88, 89, 162, 181, 202, **193.**
Gresham College, 8–11, 81.
Gresham, Sir Thomas, 8.

Haak, Theodore, 8, **194.**
Halley, E., 1, 2, 5, 88, 126, 128, 152, 190, **194.**
Harmonical Progressions, 70.
Harriot, Thos., 2, 3, 6, 7, 134–162, *passim*, 165, 183, 208, 218, 222–3, **195.**
Harvey, Wm., 2, 189, 192.
Hearne, Thos., 3, 4, 6, 178, 193, **197.**
Hérigone, 23, 151, **198.**
Heuraet, Jan, 81–82, **198.**
Hevelius, J., 127–130, 152, 194, 201, **198.**
Hobbes, Thos., 24, 65, 71, 84, 87, 166–172, 180, 209, 212, 214, **199.**
Holbech, M., 3.

238 INDEX

Holder, William, 9, 85–87, 176, **199**.
Hooke, Robt., 2, 127–130, 152, 170, 176, 194, 206, 210, 216, **200**.
Horrocks, Jeremiah, 5, 129, 180, **201**.
Huygens, 2, 7, 81–82, 99, 120, 121, 152, 155, 162, 170, 193, 210, **201**.

Imaginary Roots, 136, 142–4, 156, 158, 185.
„ Quantities, 162.
Impact, 97, 102–105, 202.
Index Notation, 138, 165, 218, 219.
Indivisibles, Method of, 15, 18, 22, 74, 186, 209.
Induction, Method of, 28, 30, 75, 78, 79, 163, 164.
Inertia, 97, 100.
Infinitesimal Calculus, 26, 90.
Infinitesimals, 26, 163.
Infinity, 18, 20.
Institutio Logicæ, 131.
" Invisible College ", 9.

Kepler, J., 26, **202**.
Kersey, 147, **203**.
" King's Cabinet ", 83.

La Géométrie, 133, 136–7, 149–51, 161.
Lagrange, J. J., 113.
Lalouère, 154, **203**.
Langbaine, Gerard, 66, 83, **203**.
Leibniz, 131, 152, 173.
Leotaud, 66, **203**.
Lloyd, Dr. William, 132.
Logarithms, 184, 194, 206.
Lucas, Henry, 5.

Mathesis Universalis, 66–71.
Mechanica, 89–90.
Mercator, 63, **204**.
Merrett, Christopher, 8, **204**.
Mersenne, Marin, 96, 97 n, 101, 185, 199, 209, **204**.
Moment of a Force, 113.
Moments, Principle of, 96.
Momentum, 103, 109–111.
Moray, Robt., 87, **205**.
Morland, Samuel, 155, **205**.
Motion, Descartes' Laws of, 97–98.
Movat, James, 3.
Mydorge, C., 23, 97, **205**.

Napier, J., 69, 142, **206**.

INDEX 239

Negative Indices, 34, 35, 40–3, 69.
,, Roots, 136, 141, 142, 143, 144, 158, 185, 191–2.
Neile, Wm., 5, 64, 80, 81–2, 84, 100, 198, **206.**
Newton, Isaac, 2–5, 28, 35, 62, 66, 88, 89, 96, 97, 101, 105–6, 131, 134, 152, 162, 170, 173, 180, 187, 190, 207, 216.
Nicholas of Cusa, 91.
Notation, 67, 218, &c.

Oldsworth, Dr., 6.
Oldenburg, Henry, 12, 81–2, 86, 88, 89, 99, 100, 104–5, 121, 152, 165, 169, 170, 188, 190, **206.**
Oresmes, 35.
Oughtred, William, 5, 13, 23, 24, 67, 68, 69, 71, 76 n, 88, 106, 138, 145, 146–8, 161, 187, 214, 219–221, **207.**
Oxford Philosophical Society, 9, 14, 99.

Paccioli, 134.
Parallelogram of Velocities, 124.
Pascal, B., 2, 5, 19, 26, 61, 73, 84, 118, 135, 154, 155.
Peletarius (Peletier), J., 65, 66, 141, **208.**
Pell, John, 5, 187, **208.**
Pell's Equation, 208.
Pepys, Samuel, 11, 129, 174, 179.
Petty, Sir Wm., 10, **209.**
Popham, Alexander, 85–86.
" Privative Roots ", 136, 142.
Projectile, Path of, 125.
Pyramidal Numbers, 51, 53.

Quadrature of the Circle, 15–16, 27, 30, 46–50, 185.

Rahn, 23, 208, **209.**
Rectification, 17, 63, 80, 81, 84, 206.
Roberval, 15, 19, 26, 27, 28, 96, 150, 151, 152, 154, 213, **209.**
Rooke, Lawrence, 5, 17, **210.**
Royal Society, 7, 8–14, and *passim*.

Saint-Vincent, Grégoire, de, 201, 203, **210.**
Savile, Sir Henry, 5, 83, 184.
Scholasticism, 4.
van Schooten, F., 15, 82, 182, 214, 215, **211.**
Sexagesimal Notation, 69.
Sloane, Sir Hans, 132, **211.**
Slusius (De Sluze), 176, **211.**
Smith, John, 13.
Smith, Dr. Robert, 3, 87, 175, **211.**
Sprat, Bishop, 11, 202.

Stevinus, Simon, 35, 93, 94–5, 98, 110, 114, 184, 191, 195, **212**.
Stifel, 67, 187.
Stubbe, Henry, 83, **212**.
Symbolic Notation, 23, 207, 214, 218, 219.

Tanner, Thos., 180, **212**.
Tartaglia, 91, 92.
Telescopic Sights, 127.
Theory of Numbers, 73, 74, 75, 79, 191, 204.
Thomson, Thomas, 136, 154.
Tides, 89.
Torricelli, 15, 27, 28, 76, 95, 99, 151, 152, 209, **213**.
Torricellian Experiment, 102.
Trigonometry, 165, 186, 192, 207.
Triangular Numbers, 50, 51, 53.
Turner, Dr. Peter, 14, **213**.
Twysden, Dr., 137.

Ubaldo del Monte, Guido, 93, **213**.
Undulatory Theory of Light, 202.

Varignon, 46.
Varro, M., 94, **213**.
Vere, Lord Horatio, 6.
Vieta, 67, 78, 133, 134, 137, 140, 146, 147, 148, 149, 150, 158–160, 161, 163, 165, 180, 195, **214**.
da Vinci, Leonardo, 92 n.
Virtual Displacements, Principle of, 107.
Viviani, 151, **214**.
Vortices, Theory of, 96.

Waller, Sir Richard, 6, 173.
Ward, Seth, 5, 10, 17, 27, 177, 188, 193, 194, 206, 208, 210, 213, **214**.
Wellar, 5.
Whaley, Daniel, 85.
White, Thomas, 71, 74.
Wilkins, Dr. John, 8, 9, 10, 66, 87, 177, 188, 189, 205, 206, 210, **215**.
Wilkinson, Henry, 66, **216**.
de Witt, Jan, 24, **215**.
Wood, Anthony à, 83, 86, 180, 189, 192, 196, 203, 212, 213, 215, 217, **216**.
Wren, Christopher, 25, 27, 64, 81–82, 99, 170, 179, 199, **216**.

Zouch, Richard, 83, **217**.

RETURN Astronomy/Mathematics/Statistics/Computer Science Library
TO ➡ 100 Evans Hall   642-3381

LOAN PERIOD 1 **7 DAYS**	2	3
4	5	6

ALL BOOKS MAY BE RECALLED AFTER 7 DAYS

DUE AS STAMPED BELOW

~~JUL 27 1982~~	APR 08 1997	
~~SEP 21 1983~~		
~~MAR 27 1984~~	APR 0 4 2005 SENT ON ILL	
~~APR 03 1984~~ ~~APR 10 1984~~	MAR 1 9 2007 U.C. BERKELEY	
~~APR 17 1984~~		
~~JUL 31 1986~~		
~~SEP 30 1991~~		
~~SEP 21 1992~~ APR 0 9 1993		

FORM NO. DD3, 10m, 11/78   UNIVERSITY OF CALIFORNIA, BERKELEY
BERKELEY, CA 94720

AMS